Lecture Notes in Computer Science 14937

Founding Editors

Gerhard Goos
Juris Hartmanis

Editorial Board Members

Elisa Bertino, *Purdue University, West Lafayette, IN, USA*
Wen Gao, *Peking University, Beijing, China*
Bernhard Steffen , *TU Dortmund University, Dortmund, Germany*
Moti Yung , *Columbia University, New York, NY, USA*

The series Lecture Notes in Computer Science (LNCS), including its subseries Lecture Notes in Artificial Intelligence (LNAI) and Lecture Notes in Bioinformatics (LNBI), has established itself as a medium for the publication of new developments in computer science and information technology research, teaching, and education.

LNCS enjoys close cooperation with the computer science R & D community, the series counts many renowned academics among its volume editors and paper authors, and collaborates with prestigious societies. Its mission is to serve this international community by providing an invaluable service, mainly focused on the publication of conference and workshop proceedings and postproceedings. LNCS commenced publication in 1973.

Apostolos Ampatzoglou · Jennifer Pérez ·
Barbora Buhnova · Valentina Lenarduzzi ·
Colin C. Venters · Uwe Zdun · Khalil Drira ·
Luciana Rebelo · Daniele Di Pompeo ·
Michele Tucci · Elisa Yumi Nakagawa ·
Elena Navarro
Editors

Software Architecture

ECSA 2024 Tracks and Workshops

Luxembourg City, Luxembourg, September 3–6, 2024
Proceedings

Editors
Apostolos Ampatzoglou
University of Macedonia
Thessaloniki, Greece

Barbora Buhnova
Masaryk University
Brno, Czech Republic

Colin C. Venters
University of Huddersfield
Huddersfield, UK

Khalil Drira
Université de Toulouse
Toulouse, France

Daniele Di Pompeo
University of L'Aquila
L'Aquila, Italy

Elisa Yumi Nakagawa
University of São Paulo
São Carlos, Brazil

Jennifer Pérez
Universidad Politécnica de Madrid
Madrid, Spain

Valentina Lenarduzzi
University of Oulu
Oulu, Finland

Uwe Zdun
University of Vienna
Vienna, Austria

Luciana Rebelo
Gran Sasso Science Institute
L'Aquila, Italy

Michele Tucci
University of L'Aquila
L'Aquila, Italy

Elena Navarro
University of Castilla-La Mancha
Albacete, Spain

ISSN 0302-9743   ISSN 1611-3349 (electronic)
Lecture Notes in Computer Science
ISBN 978-3-031-70945-6   ISBN 978-3-031-71246-3 (eBook)
https://doi.org/10.1007/978-3-031-71246-3

© The Editor(s) (if applicable) and The Author(s), under exclusive license
to Springer Nature Switzerland AG 2024

This work is subject to copyright. All rights are solely and exclusively licensed by the Publisher, whether the whole or part of the material is concerned, specifically the rights of translation, reprinting, reuse of illustrations, recitation, broadcasting, reproduction on microfilms or in any other physical way, and transmission or information storage and retrieval, electronic adaptation, computer software, or by similar or dissimilar methodology now known or hereafter developed.
The use of general descriptive names, registered names, trademarks, service marks, etc. in this publication does not imply, even in the absence of a specific statement, that such names are exempt from the relevant protective laws and regulations and therefore free for general use.
The publisher, the authors and the editors are safe to assume that the advice and information in this book are believed to be true and accurate at the date of publication. Neither the publisher nor the authors or the editors give a warranty, expressed or implied, with respect to the material contained herein or for any errors or omissions that may have been made. The publisher remains neutral with regard to jurisdictional claims in published maps and institutional affiliations.

This Springer imprint is published by the registered company Springer Nature Switzerland AG
The registered company address is: Gewerbestrasse 11, 6330 Cham, Switzerland

If disposing of this product, please recycle the paper.

Preface

The European Conference on Software Architecture (ECSA) is the premier European conference aimed at bringing together leading researchers and practitioners to present and discuss the most recent, innovative, and significant findings and experiences in the field of software architecture research and practice.

This volume complements the proceedings of the 18th edition of ECSA. It consists of the refereed and accepted papers from the Tools & Demos and Doctoral Symposium tracks of the main conference. In addition, it contains the proceedings of the two workshops co-located with the conference: the 7th Context-Aware, Autonomous and Smart Architectures International Workshop (CASA) and the 3rd International Workshop on Quality in Software Architecture (QUALIFIER). A more detailed summary of these workshops and their accepted papers is provided below. The abstract of the accepted tutorial on virtual engineering techniques for CI/CD pipelines that completed the program of the co-located events is also included in the front matter of this volume.

In total, we received 26 submissions through the EasyChair system. From these submissions, and after a single-blind review process, we accepted 15 papers for inclusion in this volume, 6 full and 9 short. Each paper was reviewed by at least 3 reviewers. We would like to thank the reviewers for both tracks and workshops for their contributions to this process.

We would also like to acknowledge the prompt and professional support from Springer, which publishes these proceedings as part of the Lecture Notes in Computer Science series. Last but not least, we would like to thank the authors of all these submissions for their contributions.

July 2024

Apostolos Ampatzoglou
Jennifer Pérez
Barbora Buhnova
Valentina Lenarduzzi
Colin C. Venters
Uwe Zdun
Elisa Yumi Nakagawa
Elena Navarro

CASA Workshop

The Context-aware, Autonomous, and Smart Architectures International Workshop (CASA) aims to address the issues and challenges raised by the design, implementation, and evaluation of today's software systems, which present characteristics such as being context-aware, dynamic, autonomous, smart, adaptive, and self-managed. The workshop brings together researchers and practitioners to discuss and exchange results, experiences, and visionary ideas concerning the above-mentioned characteristics.

CASA 2024 was the workshop's 7th edition. Previously, two special issues in the Information and Software Technology Journal have focused on the topics central to the CASA workshop.

This year, we received 8 submissions, from which 3 were accepted as full papers, achieving an acceptance rate of 38%. Each submission was evaluated by at least three PC members, based on their originality, technical quality, and adequacy to the workshop's scope. The CASA 2024 Program Committee (PC) was composed of 14 active researchers, working in several different countries of Europe.

We express our deepest gratitude to all the PC members, whose efforts were crucial in ensuring the quality of the accepted papers, and to the CASA 2024 Steering Committee for their support and contribution.

The Workshop Organizers
Khalil Drira
Luciana Rebelo

Summary of the Papers

The 2024 edition of the CASA workshop accepted 3 papers focusing on topics of using AI in enhancing smart homes, the use of blockchain technology and machine learning to control product quality, and a logistics platform for passenger and product transportation.

The paper entitled "Designing, Implementing, and Testing AI-Oriented Smart Home Applications: Challenges and Best Practices," authored by Denivan Campos et al., provides an overview of the software engineering issues in the development of AI-based solutions for smart homes. The case study presented is an academy-industry collaboration. The first contribution identifies the main challenges in developing smart homes and the essential practices adopted. The second one proposes a set of guidelines to solve the identified challenges.

The second work, entitled "Development of blockchain network for quality and product safety control information system" and authored by Aneta Poniszewska-Maranda et al., introduces the SALUS system and its architecture. SALUS is a system combining blockchain and AI to handle customers' reviews of a product, double quality of product, and their analysis from labs. The main contribution is the integration of blockchain solutions as a data supply chain in a trustworthy way, without the secure guarantee of a third party, in a distributed ledger.

Finally, the work entitled "Party without a cake? Onto an inter-modal HitchHike logistics platform for passengers and products transportation," authored by Mohammed Fahad Ali et al., suggests a novel architecture to support multi-modal logistics and mitigate transportation issues during delivery. The platform is designed to work with multiple service providers, allowing for re-planning and adapting to routing changes to maximize efficiency and reliability. The proposed solution uses an AI-based real-time planning algorithm to provide efficient delivery routes.

Summary of the Report

QUALIFIER Workshop

The QUALIFIER workshop focuses on quality aspects of software architecture. Software architectures offer a valuable opportunity to assess and drive software quality from the early stages of development through to later stages of software evolution. The objectives of this workshop are to bring together researchers and industrial practitioners from the software architecture and broader software engineering communities to collaborate, share experiences, provide directions for future research, and encourage the use of quality assessment techniques at any stage of the software engineering lifecycle.

QUALIFIER 2024 was the workshop's third edition. This year, we received 4 submissions, and after a rigorous review process, we accepted 3 outstanding papers that contribute to various aspects of software architecture and quality assessment. Each submission was evaluated by at least three PC members, based on its originality, technical quality, and relevance to the workshop's scope.

We express our deepest gratitude to the Program Committee for their dedication and effort in ensuring the high quality of the selected papers.

The Workshop Organizers
Daniele Di Pompeo
Michele Tucci

Summary of the Papers

The 2024 edition of the QUALIFIER workshop accepted 3 papers focusing on innovative approaches to leveraging large language models for architecture refactoring, mining software repositories for sustainability in cloud architecture, and understanding the positive side effects of software architecture evaluation.

The paper entitled "Mapping source code to software architecture by leveraging Large Language Models," authored by Nils Johansson et al., addresses the challenge of architecture refactoring, which requires thorough analysis and labor-intensive activities to restructure functionalities from a legacy architecture to a new intended one. The authors propose a novel approach to automatically map source code to the intended architecture by leveraging large language models. By formulating the problem as a machine learning text classification task, the methods are mapped into architectural modules using various approaches. The study demonstrates that vectorizing text and code using large language models outperforms other modern methods, with the best machine learning classifiers achieving around 40% accuracy and 30% F1-score.

The paper entitled "Mining for sustainability in cloud architecture among the discussions of software practitioners: building a dataset," authored by Sahar Ahmadisakha et al., addresses the increasing adoption of cloud computing in designing and implementing software systems and the necessity of considering sustainability implications. This paper bridges the gap between academic research and practitioner perspectives by using software repository mining techniques to analyze 192 discussions among practitioners on the StackExchange platform. The goal is to build an annotated dataset of cloud architectural discussions and understand the current discourse on sustainability in cloud architecture. Initial findings reveal that practitioners focus on design aspects—analysis, synthesis, and implementation—while avoiding complex activities like evaluation and maintenance. The study highlights a contrast with previous literature by emphasizing technical sustainability and economic dimensions over environmental sustainability.

The paper entitled "Positive Side-Effects of Evaluating a Software Architecture," authored by Pablo Cruz et al., presents a series of positive side effects—unintended beneficial consequences—of evaluating software architectures. The authors describe various positive side effects observed in several architecture evaluations they have led. These include enhanced stakeholder communication, improved design decisions, and increased awareness of architectural issues. The paper provides detailed descriptions of these effects and the circumstances under which they were observed, encouraging practitioners to adopt software architecture evaluation and prompting researchers to explore new research issues in this domain.

Organization

General Co-chairs

Elena Navarro — University of Castilla-La Mancha, Spain
Elisa Yumi Nakagawa — University of São Paulo, Brazil

Steering Committee

Paris Avgeriou (Chair) — University of Groningen, The Netherlands
Thais Batista — Federal University of Rio Grande do Norte, Brazil
Stefan Biffl — Vienna University of Technology, Austria
Tomas Bures — Charles University, Czechia
Carlos E. Cuesta — Rey Juan Carlos University, Spain
Laurence Duchien — University of Lille, France
Matthias Galster — University of Canterbury, New Zealand
Ilias Gerostathopoulos — Vrije Universiteit Amsterdam, The Netherlands
Paola Inverardi — University of L'Aquila, Italy
Patricia Lago — Vrije Universiteit Amsterdam, The Netherlands
Grace Lewis — Carnegie Mellon Software Engineering Institute, USA
Ivano Malavolta — Vrije Universiteit Amsterdam, The Netherlands
Raffaela Mirandola — Politecnico di Milano, Italy
Henry Muccini — University of L'Aquila, Italy
Elisa Yumi Nakagawa — University of São Paulo, Brazil
Elena Navarro — University of Castilla-La Mancha, Spain
Flavio Oquendo — IRISA, University of South Brittany, France
Ipek Ozkaya — Carnegie Mellon University, USA
Patrizia Scandurra — University of Bergamo, Italy
Bedir Tekinerdogan — Wageningen University, The Netherlands
Chouki Tibermacine — University of Montpellier, France
Catia Trubiani — Gran Sasso Science Institute, Italy
Danny Weyns — KU Leuven, Belgium

Tools and Demos Co-chairs

Apostolos Ampatzoglou — University of Macedonia, Greece
Jennifer Perez — Universidad Politécnica de Madrid, Spain

Tools and Demos Program Committee

Antinisca Di Marco — University of L'Aquila, Italy
Jasmin Jahic — University of Cambridge, UK
Angelika Musil — Technische Universität Wien, Austria
Nour Ali — Brunel University London, UK
Sara Hassan — Birmingham City University, UK
Henry Muccini — University of L'Aquila, Italy
Dalia Sobhy — AASTMT, Egypt
Rami Bahsoon — University of Birmingham, UK
André van Hoorn† — University of Hamburg, Germany
Abdessalam Elhabbash — Lancaster University, UK

Doctoral Symposium Co-chairs

Barbora Buhnova — Masaryk University, Czechia
Valentina Lenarduzzi — University of Oulu, Finland

Doctoral Symposium Program Committee

Steffen Becker — University of Stuttgart, Germany
Tomas Bures — Charles University, Czechia
Jan Carlson — Mälardalen University, Sweden
Andrea Janes — Free University of Bozen-Bolzano, Italy
Anne Koziolek — Karlsruhe Institute of Technology, Germany
Raffaela Mirandola — Karlsruhe Institute of Technology, Germany
Henry Muccini — University of L'Aquila, Italy
Davide Taibi — University of Oulu and Tampere University, Finland
Catia Trubiani — Gran Sasso Science Institute, Italy
Uwe Zdun — University of Vienna, Austria

Workshops and Tutorials Co-chairs

Colin C. Venters — University of Huddersfield, UK
Uwe Zdun — University of Vienna, Austria

Context-Aware, Autonomous and Smart Architecture (CASA) Workshop

Workshop Co-chairs

Khalil Drira — Université de Toulouse, France
Luciana Rebelo — Gran Sasso Science Institute, Italy

Workshop Program Committee

Michel Albonico — Universidade Tecnológica Federal do Paraná, Brazil
Ricardo Caldas — Chalmers University of Technology, Sweden
Rafael Capilla — Universidad Rey Juan Carlos, Spain
Mauro Caporuscio — Linnaeus University, Sweden
Amleto Di Salle — Gran Sasso Science Institute, Italy
Liliana Dobrica — University Politehnica of Bucharest, Romania
Khalil Drira — University of Toulouse, France
Cédric Eichler — INSA Centre Val de Loire, France
Pasqualina Potena — RISE Research Institutes of Sweden AB, Sweden
Claudia Raibulet — Vrije Universiteit Amsterdam, The Netherlands
Luciana Rebelo — Gran Sasso Science Institute, Italy
Elvinia Riccobene — University of Milan, Italy
Ramon Salvador Vallès — Universitat Politècnica de Catalunya, Spain
Thierry Villemur — LAAS-CNRS, France

International Workshop on Quality in Software Architecture (QUALIFIER)

Workshop Co-chairs

Daniele Di Pompeo — University of L'Aquila, Italy
Michele Tucci — University of L'Aquila, Italy

Workshop Program Committee

Nour Ali	Brunel University London, UK
Jan Bosch	Chalmers University of Technology, Sweden
Giordano d'Aloisio	University of L'Aquila, Italy
Sebastian Frank	University of Stuttgart, Germany
Giovanni Quattrocchi	Politecnico di Milano, Italy
António Rito Silva	Universidade de Lisboa, Portugal
Jacopo Soldani	University of Pisa, Italy

Organizing Committee

Program Co-chairs

Matthias Galster	University of Canterbury, New Zealand
Patrizia Scandurra	University of Bergamo, Italy

Industrial Track Co-chairs

Pablo Oliveira Antonino	Fraunhofer IESE, Germany
Tommi Mikkonen	University of Helsinki, Finland

Journal First Co-chairs

Jesper Andersson	Linnaeus University, Sweden
Robert Heinrich	Karlsruhe Institute of Technology, Germany

MIP Co-chairs

Paris Avgeriou	University of Groningen, The Netherlands
Eoin Woods	Endava, UK

Open Science Chair

Justus Bogner	Vrije Universiteit Amsterdam, The Netherlands

Proceedings Co-chairs

Vasilios AndrikopoulosUniversity of Groningen, The Netherlands
Jamal El HachemUniversity of South Brittany, IRISA Laboratory, France

Publicity and Social Media Co-chairs

Matteo CamilliPolitecnico di Milano, Italy
Joanna C. S. SantosUniversity of Notre Dame, USA

Student Volunteer Chair

Aurora MaciasUniversity of Castilla-La Mancha, Spain

Web Chair

Wallace ManzanoUniversity of São Paulo, Brazil

Local Chair

Magali MartinUniversity of Luxembourg, Luxembourg

Virtual Engineering Techniques for CI/CD Pipelines (Tutorial Abstract)

Pablo Oliveira Antonino[1], Marius Becker[2], Priom Biswas[3], and Benedikt Lüken Winkels[4]

[1] Fraunhofer IESE, Kaiserslautern, Germany
[2] Fraunhofer IESE, Germany
[3] Fraunhofer IESE, Germany
[4] Fraunhofer IESE, Kaiserslautern, Germany

Summary

Virtual engineering techniques such as simulation and digital twins are increasingly being used in the development of dependable systems to improve the performance and accuracy of CI/CD pipelines. Seamless integration of digital models into the development process has been a key enabler for continuous and accurate analysis of design space exploration, which ultimately enhances the analysis of the impact of architecture decisions on strategic business goals. In this regard, this tutorial will provide an overview and practical sections on the technologies and methodologies developed to address these issues. Specifically, the following topics will be covered: Motivations for virtual engineering and architecture-centric continuous engineering emerging in different industries, basics of simulations and digital twins and their technological realizations possibilities and strategies to integrate them into CI/CD pipelines.

Contents

Tools and Demos

DiSpel Cockpit: Specification, Verification, and Refinement of Resilience
Scenarios 3
   *Sebastian Frank, Aref El-Maarawi Tefur, Alireza Hakamian,
and André van Hoorn*

OAS2Tree: Visual API-First Design 12
   Souhaila Serbout and Cesare Pautasso

A Multi-variant Benchmark for Microservice Systems in Software
Engineering Research 21
   *Tomas Cerny, Md Showkat Hossain Chy,
Muhmmad Ashfakur Rahman Arju, Korn Sooksatra,
Amr S. Abdelfattah, and Valentina Lenarduzzi*

Semantics Enhancing Model Transformation for Automated Constraint
Validation of Palladio Software Architecture to MontiArc Models 30
   *Sebastian Weber, Jörg Henß, Bahareh Taghavi, Thomas Weber,
Sebastian Stüber, Adrian Marin, Bernhard Rumpe, and Robert Heinrich*

Extending a Low-Code Tool with Multi-cloud Deployment Capabilities 39
   Fitash Ul Haq, Iván Alfonso, Armen Sulejmani, and Jordi Cabot

Doctoral Symposium

Reference Architecture of MLOps Workflows 49
   Faezeh Amou Najafabadi

Evaluating the Effect of Team Ownership of Microservices: Strategies
for Balancing Decoupling, Coordination, and System Cohesion 58
   Noman Ahmad

Technical Debt and Software Quality in Cloud-Native Applications 65
   Ruoyu Su

Improving QoS of Microservices Architecture Using Machine Learning
Techniques 72
   Neha Kaushik

CASA Workshop

Designing, Implementing, and Testing AI-Oriented Smart Home
Applications: Challenges and Best Practices 83
    *Denivan Campos, Luana Martins, Joselito Mota, Dhyego Tavares,
    Jander Pereira, Mayki Oliveira, Denis Boaventura, Diego Correa,
    Eduardo Ferreira, George Pinto, Nilton Seixas, Adriano Maia,
    Matias Romário, Ernando Passos, Frederico Durao,
    Gustavo B. Figueiredo, Maycon Peixoto, Tiago Januario,
    Cassio Prazeres, Ivan Machado, and Eduardo Almeida*

Party Without a Cake? Onto an Inter-modal HitchHike Logistics Platform
for Passengers and Products Transportation 100
    *Mohammed Fahad Ali, Dominique Briechle, Marit Briechle-Mathiszig,
    Tobias Geger, and Andreas Rausch*

Development of Blockchain Network for Quality and Product Safety
Control Information System 115
    *Aneta Poniszewska-Marańda, Michał Pawlak, Maciej Kopa,
    and Mateusz Owczarek*

QUALIFIER Workshop

Mapping Source Code to Software Architecture by Leveraging Large
Language Models 133
    Nils Johansson, Mauro Caporuscio, and Tobias Olsson

Mining for Sustainability in Cloud Architecture Among the Discussions
of Software Practitioners: Building a Dataset 150
    Sahar Ahmadisakha and Vasilios Andrikopoulos

Positive Side-Effects of Evaluating a Software Architecture 167
    Pablo Cruz and Hernán Astudillo

Author Index 179

Tools and Demos

DiSpel Cockpit: Specification, Verification, and Refinement of Resilience Scenarios

Sebastian Frank[✉], Aref El-Maarawi Tefur, Alireza Hakamian, and André van Hoorn

University of Hamburg, Hamburg, Germany
{sebastian.frank,aref.el-maarawi.tefur}@uni-hamburg.de

Abstract. Chaos Engineering is an established method to assess the resilience of software systems by injecting failures and learning from experiments in production. Existing Chaos Engineering tools, such as Chaos Toolkit, facilitate creating and executing various failures but lack support for the entire process of resilience scenario elicitation, specification, execution, and refinement. This paper introduces DiSpel Cockpit for continuous and iterative specification, verification, and refinement of resilience scenarios. To achieve its goal, the DiSpel Cockpit combines the capabilities of existing tools into a holistic approach.

The DiSpel Cockpit uses Property Specification Patterns as a formalism to specify stimuli and responses of scenarios. System data is obtained from simulations and monitoring data. This paper presents the tool and demonstrates its usefulness based on resilience scenarios for an industrial system. We expect DiSpel Cockpit to assist software architects, particularly in the early phases of applying Chaos Engineering, when scenarios still have to be formalized, and feedback is necessary to gain confidence before moving toward conducting experiments in production.

We provide a video[1], source code, example data, and Docker containers[2].

1 Introduction

Resilience can be defined as a system's ability to continue its operation under adverse condition and to recover [12]. It is a quality that should be taken into account particularly when designing microservice-based software systems [13]. In the industry, Chaos Engineering (CE) [2] is an established method to build confidence in a system's resilience by injecting failures in production. It comprises an experimentation process in which hypotheses are stated and refined.

In previous work [7], we successfully designed and conducted a workshop to elicit resilience requirements in the form of quality scenarios as introduced

---

[1] Demo Video: https://youtu.be/gP6USBfOuxY.
[2] Source Code & Releases: https://github.com/Cambio-Project/DiSpel-Cockpit.

by Bass et al. [3]. As the CE methodology suggests, we turned some of these scenarios into chaos experiments. Although we were eventually successful, we also experienced challenges in the process. While plenty of tooling is available for conducting chaos experiments, e.g., Chaos Toolkit [4], we found the limited tool support for other activities in the CE process challenging. However, we also found conceptual challenges in the early phase of applying CE. CE as a method does not give explicit advice on how to (i) transform elicited resilience scenarios into experiments, (ii) get quick feedback on the feasibility of the scenario specifications, and (iii) refine resilience scenarios.

The DiSpel Cockpit implements the DiSpel approach [8], which aims to solve these issues. It provides a web-based graphical user interface and coordinates the extended and containerized tools PSPWizard [1], MiSim [10], MoSIM, TBVerifier [9], and TQPropRefiner [6]. Using our tool and PSPWizard [1], software architects can specify resilience scenarios using Property Specification Patterns (PSP) [1] to obtain testable scenarios. Next, relevant scenario occurrences can be obtained through simulation with MiSim [10] and search in monitoring data with MoSIM as a quick way to get feedback without setting up experiments. Our tool then performs runtime verification [11] with TBVerifier [9] to determine whether response specifications are satisfied. Finally, the scenarios can be refined by adjusting the parameter values in the provided PSP using TQPropRefiner [6].

In contrast to established CE tooling, like Chaos Toolkit [4], our approach is scenario-based, considers simulation and monitoring as data sources, and provides refinement strategies. Our previous work TQPropRefiner [6] is similar in some aspects but allows only specification of a single (response) PSP and excludes the detection of relevant occurrence data. Thus, the contribution of this work is a holistic, scenario-based CE tool focusing on the early phases of CE.

2 DiSpel Approach

In previous work, we introduced the vision of data-driven DiSpel approach [8], which proposes a continuous and iterative process for the specification, verification, and refinement of resilience scenarios as depicted in Fig. 1. As such, the DiSpel approach is a derivative of CE [2]. In contrast to CE, the DiSpel approach considers a wider variety of data sources, i.e., besides chaos experiments, simulation, and monitoring data. Further, it introduces scenarios as a means to describe hypotheses, provides concrete refinement strategies, and allows analyses of transient behavior during experiments.

In the **specification** phase, software architects state hypotheses as scenarios [3], which consist of the elements stimulus, response, and environment, among others. PSP are used in the specification process for stimuli and responses. PSP [1] can be described as templates for common specifications. They can be represented in human-readable, Structured English Grammar (SEG) and translated into various testable, temporal logics.

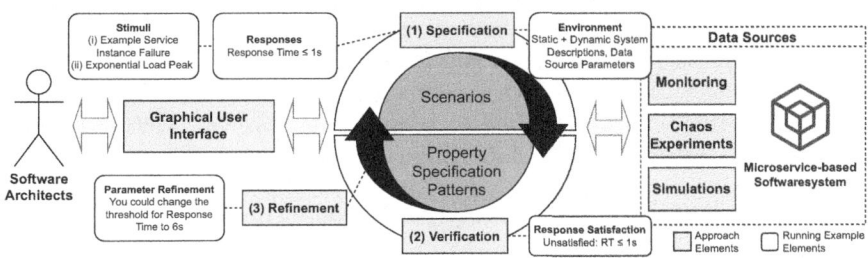

Fig. 1. Simplified depiction of the DiSpel Approach (adopted from [8])

In the **verification** phase, runtime verification [11] can be used to test against system data from various data sources since the scenarios are formally specified. The DiSpel approach considers *active* and *passive* data collection. Experiments and simulations can be actively triggered through the specifications to gather the data on the system's response. Simulations usually require modeling effort and lack precision but enable quick feedback and the potential to analyze what-if scenarios. As a passive and resource-efficient method, the specified incidents can be identified in monitoring data, if available. Note that all these methods will be applied based on the same (stimuli) specification.

In the **refinement** phase, practitioners can gain insights from the verification results on choosing feasible and appropriate resilience hypotheses for their system. By suggesting adjustments to parameter values of the response, such as a response time threshold, or modifying the entire pattern type of the response, practitioners can strengthen or weaken the response for a given scenario. This step is crucial due to the uncertainty in specifying exact values beforehand, as software architects often struggle to determine if their specifications are feasible.

3 Running Example

In a workshop, we elicited 12 resilience scenarios for a real payment accounting system under development, designed with a microservice-based architecture [7]. The sixth scenario describes an instance failure caused by a software bug and the third scenario an exponentially increasing load peak. As a running example, we synthesized a new scenario by combining these scenarios into a more complex scenario. This scenario reads as *immediately after an instance failure caused by a software bug, the number of wage clerks using the system rises exponentially during the payslip period at regular service hours. The system is expected to perform within a guaranteed tolerance, ensuring wage clerks always receive correct answers within 1 s (99% of the time).* We use this example to demonstrate the tool's workflow and main features.

In our scenario, we identify two stimuli (instance failure and load peak) and one response (response time should be below 1 s, 99% of the time). Using PSP, we formalize the response by applying the Universality pattern as follows: *Globally, it is always the case that ResponseTimeOK holds.* Here, *ResponseTimeOK* is the system state more precisely described as *response time* ≤ *1 s*. Similarly, stimuli can be described through Existence and Response patterns. The full example is provided together with the tool. Once stimuli and responses are specified using suitable PSPs, they can be mapped to temporal logics, e.g., Metric Temporal Logic (MTL) as described in the following:

$$\Box(\text{ResponseTimeOK}(\text{AllResponseTimes}))$$

We introduce the predicate ResponseTimeOK() with the AllResponseTimes metric as a parameter. The temporal operator □ means *always*. Thus, this expression evaluates to true when AllResponseTimes is less than 1 s for all its values.

4 DiSpel Cockpit

The DiSpel Cockpit serves as a unified platform that offers software architects a web-based user interface, which seamlessly integrates and coordinates existing tools, allowing our entire process to be managed and executed within the tool itself. We continue with the running example (Section 3 and Fig. 1) to demonstrate how a user can apply the DiSpel approach (Section 2) using the DiSpel Cockpit. Excerpts from the user interface are displayed in Fig. 2. We refer to the video and the example inputs on the project's GitHub page for details.

4.1 Specification

The user starts by creating a new resilience scenario on the *Scenario Editor* page. Creating a new scenario involves specifying the three main components of the scenario [3] format: the stimuli, the response, and the environment. The user can also provide a textual description of the scenario and a category tag for filtering and grouping related scenarios.

To set up the environment, the user uploads the necessary files, configuring the simulation and monitoring data retrieval processes implemented through MiSim [10] and MoSIM, respectively. The simulation requires an architectural model of the system under test, describing its architecture's static properties, like services, dependencies, and operations. Additionally, the user provides the experiment description and load profiles, which describe the dynamic properties of the system. For data retrieval from monitoring logs, users must supply monitoring data (currently a CSV file) and set a parameter for the MoSIM tool to determine the duration of the occurrences extracted from the monitoring data.

The stimuli and responses are specified using PSP. The *PSP Editor* page facilitates the creation of specifications through a UI. Users construct the specification by selecting the scope and appropriate pattern type, followed by the pattern itself and any additional building blocks compatible with the chosen pattern. Figure 2 (A) shows a part of the UI after specifying the running example's response. In the background, the PSPWizard [1] translates the specified pattern into the selected target logic, and the results are displayed and persisted.

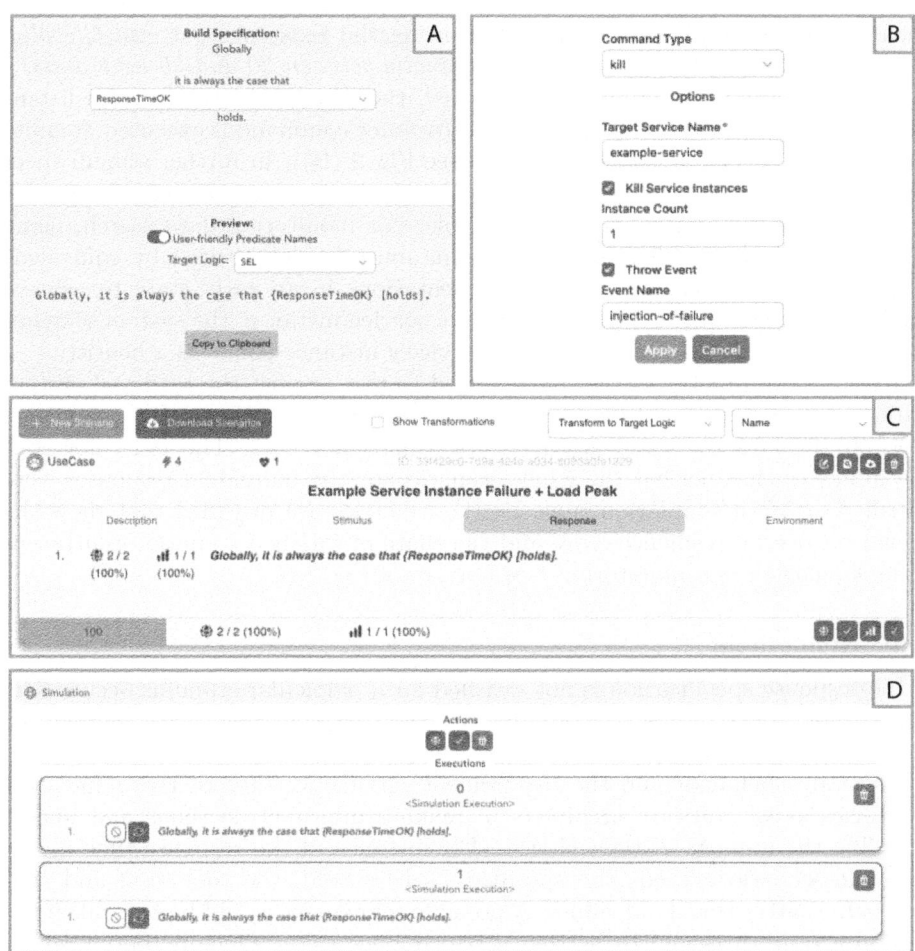

Fig. 2. Selected screenshots of the DiSpel Cockpit UI

4.2 Verification

Fig. 2 (C) shows the compact presentation of a fully specified scenario in the *Scenario Overview*, a list of all specified scenarios. The user can initiate the verification process by running simulations and searches over monitoring data. The specification of stimuli and responses is fundamentally similar, particularly regarding the pattern selection. However, responses contain only events derivable from system metrics. In contrast, stimuli can (and should) also contain commands because stimuli serve as the instructions for retrieving relevant system data. The DiSpel Cockpit assists the software architect in formulating listenable events and commands by providing wizards. Figure 2 (B) shows the wizard for specifying the *kill* command *killExampleServiceInstance*. This command can then be used in a stimulus specification using the Existence pattern: *Globally, (killExampleServiceInstance) [holds] eventually between 20 and 20 time units.*

Note that once a command is executed, the simulator can trigger a listenable event. After the *killExample-Service-Instance* command is executed, it emits the listenable event *injection-of-failure* (see Fig. 2 (B)). In further stimuli specifications, this event acts as a trigger to induce a workload peak. Currently, only *kill* and *load* commands are available. For monitoring data search, using the same stimuli specifications, commands must be substituted by equivalent events. MoSIM provides default implementations in an early state to achieve this. For example, instead of terminating a service instance, the system searches the monitoring data for a drop in the service's instance count as a heuristic.

Responses are later in the process used to test against the retrieved system data, which is performed by TBVerifier [9]. As shown in Fig. 2 (D), the Dispel Cockpit displays the verification results using color cues. The red color coding indicates that neither the whole scenario nor the individual responses were satisfied for both simulation runs. Further, metrics are provided that show the scenario's overall resilience score and the share of satisfied scenario occurrences from simulation and monitoring (see bottom of Fig. 2 (C)).

4.3 Refinement

If the response specification is not satisfied for a particular stimulus occurrence, the Dispel Cockpit aids the user in investigating the cause and refining the specification. The *Refinement View*, powered by TQPropRefiner [6], visualizes the system's behavior and the requirement satisfaction. Interactive refinement strategies assist software architects in making informed decisions and understanding the impacts of their choices. For instance, if the response time in our example scenario exceeds the specified 1 s threshold, the tool tests and suggests alternative threshold values. A possible refinement would be to weaken the response time threshold by setting it to a higher value. Once potential refinements are identified and accepted, a new cycle of verification and re-refinement can begin.

5 Implementation

The DiSpel Cockpit is a web-based application following a client-server model. The frontend is a single-page application built with the lightweight Nuxt.js, a Vue.js-based framework. In contrast to Vue, Nuxt enables backend development support, allowing communication between the frontend and backend services. MongoDB is responsible for persisting the scenarios and analysis metrics. System data is currently stored in a shared Docker volume.

The Cockpit's backend is structured as microservices architecture, developed using different programming languages (Java, Kotlin, Python) and frameworks (Angular), integrating the five tools PSPWizard [1], MiSim [10], MoSIM, TBVerifier [9], and TQPropRefiner [6] as services. These services run in Docker containers orchestrated using Docker Compose, enabling independent deployment and enhancing flexibility and scalability.

Most backend services previously lacked APIs and provided GUIs only. For PSPWizard and MiSim, we implemented REST APIs by encapsulating the API functionality within a Java submodule using the Spring Boot framework to retain these tools' standalone capability. TQPropRefiner required adaptions of functionality, GUI components, and its API for integration into our Vue frontend. We established the TQPropRefiner as a separate service, embedding its GUI elements within the Vue.js frontend as a web page. This facilitated seamless integration without extensive code modification. Some (functional) service changes were necessary to allow for proper interaction between the tools. Among others, we added a modern UI and a new target language to the PSPWizard to match the syntax expected by TBVerifier and extended TQPropRefiner to generate its GUI dynamically for support of various PSP types.

6 Discussion

During the development of the DiSpel Cockpit and its underlying tools, we have been in exchange with an industry partner to ensure the tool's usefulness for practical use cases. In early feedback, practitioners praised the expressiveness of the PSP-based resilience scenarios and the capabilities for analyzing transient behavior. In addition, the running example used during development is based on real scenarios [7], and the simulation model has been reused from previous work [10]. Further, PSP have shown to be sufficient to capture quality requirements in industrial case studies [1]. With limited generalizability, this indicates the feasibility of our approach. Nevertheless, a more thorough and systematic evaluation of the tool's capabilities to aid software architects in the early phases of CE is still necessary.

While there is no evaluation of the DiSpel approach and tool as such, specific aspects and underlying tools have been partially evaluated before [6,9,10]. For example, expert users solving tasks with TQPropRefiner were mostly successful and found it easy to refine specifications [6]. Further, Czepa and Zdun [5] have shown the understandability of PSP be superior to plain temporal logic.

The DiSpel Cockpit currently suffers from technical and conceptual limitations, both partially originating from the underlying tools. While conceptually compatible with PSP, the tooling's implementation currently does not support composed and complex predicates (like *Service 1 fails and Service 2 fails*) and the handling of time units. It also requires equally sized time steps in the analyzed data. Further, the supported command (currently: kill and load), listener (currently: user defined events), and PSP types must be extended. Regarding the conceptual limitations, early user feedback suggests collecting monitoring data from monitoring systems, executing actual chaos experiments, and adding support in eliciting and (graphically) specifying scenarios would increase the tool's utility. By addressing these limitations, we plan to extend the work to support more (realistic) scenarios and industrial, real-world systems.

7 Conclusion

In this paper, we presented the DiSpel Cockpit, a tool that allows software architects to specify, verify, and refine resilience scenarios. To reach this goal, the DiSpel Cockpit leverages the capabilities of existing tools, i.e., the PSPWizard for specifying PSP, the resilience simulator MiSim and the stimuli search library MoSIM for generating/finding data, the TBVerifier for analyzing scenario satisfaction, and the TQPropRefiner for refining scenarios.

Although we provided an essential proof-of-concept of the DiSpel approach through the DiSpel Cockpit, a thorough evaluation of its usefulness and presumed benefits in a more realistic setting is still necessary, e.g., in user studies. In future work, we also intend to improve and extend the connected tools and the DiSpel Cockpit itself, e.g., by adding transformations to chaos experiments, employing graphical specification, and assisting in eliciting scenarios.

Acknowledgments. The authors thank Angelina Heinrichs, Marvin Taube, Alexander Baur, Patrick Mohr and Julian Brott for contributions to the tool and the German Federal Ministry of Education and Research (dqualizer FKZ: 01IS22007B and Software Campus 2.0 — Microproject: DiSpel, FKZ: 01IS17051) for supporting this work.

Disclosure of Interests. The authors have no competing interests to declare that are relevant to the content of this article.

References

1. Autili, M., Grunske, L., Lumpe, M., Pelliccione, P., Tang, A.: Aligning qualitative, real-time, and probabilistic property specification patterns using a structured English grammar. IEEE Trans. Software Eng. **41**(7), 620–638 (2015)
2. Basiri, A., et al.: Chaos engineering. IEEE Software **33**, 1–1 (01 2016)
3. Bass, L., Clements, P., Kazman, R.: Software architecture in practice. Addison-Wesley Professional, 4 edn. (2021)
4. Chaos Toolkit Team: Chaos Toolkit (2023). https://chaostoolkit.org
5. Czepa, C., Zdun, U.: On the understandability of temporal properties formalized in linear temporal logic, property specification patterns and event processing language. IEEE Trans. Software Eng. **46**(1), 100–112 (2018)
6. Frank, S., Brott, J., Hakamian, A., van Hoorn, A.: TQPropRefiner: interactive comprehension and refinement of specifications on transient software quality properties. In: ECSA'23 Post-Proceedings (2023), (in press)
7. Frank, S., Hakamian, A., Wagner, L., Kesim, D., Zorn, C., von Kistowski, J., van Hoorn, A.: Interactive elicitation of resilience scenarios based on hazard analysis techniques. In: ECSA'21 Post-Proceedings, pp. 229–253. Springer (2021). https://doi.org/10.1007/978-3-031-15116-3_11
8. Frank, S., Hakamian, A., Wagner, L., von Kistowski, J., van Hoorn, A.: Towards continuous and data-driven specification and verification of resilience scenarios. In: ISSREW'22, pp. 136–137. IEEE (2022)
9. Frank, S., Hakamian, A., Zahariev, D., van Hoorn, A.: Verifying transient behavior specifications in chaos engineering using metric temporal logic and property specification patterns. In: ICPE'23 Companion, p. 319-326. ACM (2023)

10. Frank, S., Wagner, L., Hakamian, A., Straesser, M., van Hoorn, A.: MiSim: A simulator for resilience assessment of microservice-based architectures. In: QRS'22, pp. 1014–1025. IEEE (2022)
11. Leucker, M., Schallhart, C.: A brief account of runtime verification. J. Logic Algebraic Program. **78**(5), 293–303 (2009)
12. National Institute of Standards and Technology: NIST SP 800-39 (2011)
13. Newman, S.: Building Microservices. O'Reilly Media (2015)

OAS2Tree: Visual API-First Design

Souhaila Serbout(✉) and Cesare Pautasso(✉)

Software Institute (USI), Lugano, Switzerland
{souhaila.serbout,cesare.pautasso}@usi.ch

Abstract. OAS2tree is a tool designed to transform OpenAPI Specification (OAS) documents into tree-like visualizations, aiding in the understanding and navigation of the structure of REST APIs. By converting the detailed, verbose, and often complex OAS files into a visual tree structure, OAS2tree simplifies the comprehension of a Web API, highlighting the hierarchical relationships between endpoints, operations, and parameters. This visual representation is particularly useful for developers and stakeholders who need a quick overview of an API without delving into the intricate details of its technical specifications. OAS2tree can be integrated into the IDE through a Visual Studio code extension or used as a standalone web application. The tool currently has about 400 users and has been used on teaching, research, and development projects. In this paper, we present the design and implementation of OAS2tree, highlighting its features and use cases. We also highlight the limitations of the current version and discuss future improvements and potential extensions.

Demo Video Link: https://youtu.be/E48c9Rwntz8

1 Introduction

Web APIs are essential for enabling seamless communication between software systems [10]. However, their sheer size and complexity [7] can make them challenging to understand [4]. While visualizing the overall architecture can significantly aid in understanding the interactions between a system's components, focused visualizations of crucial elements within a complex system, such as web APIs, can clarify the structure, help ensure the correct flow of data, and determine which operations should be exposed.

OAS2Tree dynamically generates visual representations of APIs described using the OpenAPI Specification (OAS) [5]. The tree-like visualization represents the API endpoints, request/response elements, and parameters. By offering developers a visual representation instantaneously synchronized from the OpenAPI description text, the tool facilitates obtaining valuable insights and a deeper understanding of the API structure. Furthermore, our tool goes beyond basic visualization capabilities by proactively identifying and highlighting design smells—common issues or inefficiencies in API design. These design smells have been extensively discussed in our previous research study on API structural patterns and design flaws [9].

In this paper, we outline the main features of OAS2tree. By offering developers the ability to interact with visual representations and detect design smells early on, the tool empowers developers to ensure they create the wanted API and consumers to inspect and select their chosen API. By providing real-time visual representations of API endpoint and operation structures, the tool facilitates validating the consistent and regular design of API endpoints as a tree of URL paths with color-coded HTTP methods as tree leaves. Additionally, its integration of design smell detection capabilities assists developers in identifying potential issues and refining their APIs. Ultimately, our visualization tool aims to enhance the collaboration and efficiency of incremental API design and review processes, resulting in the creation of well-designed APIs.

The tool is available as both a standalone web app [1] and as a Visual Studio Code (VSCode) extension [2], catering to the diverse needs of developers. The web app version is designed for those who want a quick and convenient way to visualize an API without the need for local tool installation and setup. On the other hand, the VSCode extension version is intended for developers who prefer to have the visualization tool seamlessly integrated into their development environment.

2 From OpenAPI to API Tree

To visualize the OpenAPI specification as a tree structure, we transform the flat list of paths extracted from an API description into a hierarchical structure by breaking down the paths into segments. When a segment is shared between different paths, we check if these segments share the same sequence of parent segments. If they do, the segments are merged into a single node, as they refer to the same container resource. Once the path segments tree structure is complete we attach to each node the set of HTTP methods of the corresponding endpoint.

OpenAPI also includes details related to the responses of each endpoint and the parameters that can be passed to the endpoint. We attach these details to the corresponding HTTP method node. In the current version of OAS2Tree, we only visualize the response status codes but do not further drill down to show the data models of the request or response schemas.

Considering the example in Listing 1.1 of API paths defined in an OpenAPI document:

Listing 1.1. Example of API paths description ins an OpenAPI document

```
paths:
  /users:
    get:
      summary: "Retrieve the list  of all registered users."
      responses:
        "200":
          description: "A list of users."
  /users/{id}/details:
    get:
```

```
        summary: "Retrieve details of a user by their ID."
        responses:
          "200":
            description: "The user information."
          "404":
            description: "User not found."
  /posts:
    get:
      summary: "List all posts"
      responses:
        "200":
          description: "A list of all available posts."
          parameters:
          - name: limit
            in: query
            description: "The number of posts to return."
            type: integer
          - name: offset
            in: query
            description: "The number of posts to skip."
            type: integer
    post:
      summary: "Create a new post."
      responses:
        "201":
          description: "The created post."
```

The paths /users and /users/{id}/details share the segment /users, and the paths /users/{id}/details and /posts share the segment /users/{id}. Since these segments share the same sequence of parent segments, they are merged into one node in the tree structure. The resulting tree structure can be visualized as in Fig. 1.

While the YAML or JSON syntax adopted by OpenAPI can make it hard to locate the operations that are applicable over the same resources but with

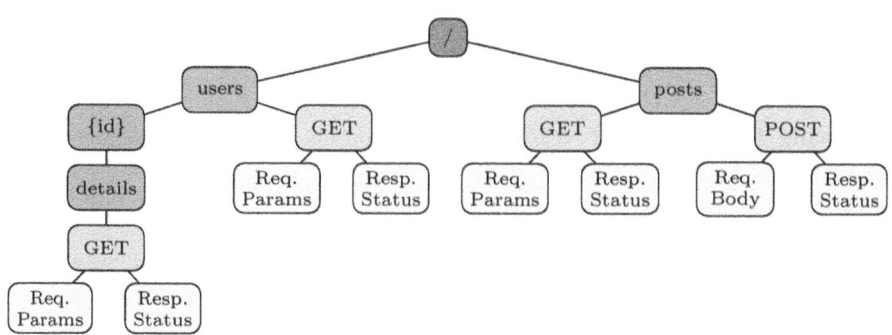

Fig. 1. Extracted API tree structure

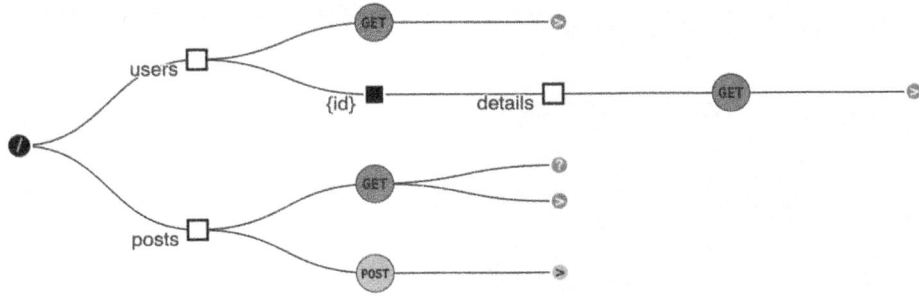

Fig. 2. OAS2Tree visualization of API in Listing 1.1

different HTTP methods, in case of large APIs with many paths, the tree structure makes it easy to see that – for example – the /users/{id}/details path has only a GET operation and the /posts path has both a GET and a POST operation. This hierarchical representation clearly shows the shared segments and their relationships within the API structure.

3 OAS2Tree Elements Graphical Notation

We defined the following set of notations to represent the different elements of the API tree structure (Fig. 2).

HTTP Methods: They are visualized in a circular shape. We attribute a specific color to each HTTP method to make it easier to identify them in the visual representation. The colors we adopted are closely similar to the color coding used in Postman [6]. The colors are as follows:

Path segments: we distinguish the position of fixed URL segments (represented with a white tree node □) from the position of in-path parameters (represented with a black tree node ■). We also use a different notation ( ) to distinguish unusual path segments. For example in /api/v1/users/{userId}:posts the last suffix of the path ":posts" is prefixed by a colon character as opposed to the usual forward slash.

Query parameters: we represent query parameters as a subtree of the operation node. The subtree is collapsed by default, to highlight the presence of query parameters, and the user can expand it to see which parameters are expected by the operation. The root of the subtree is represented by a question mark icon ⓘ . Then each parameter is represented by a node with the parameter name (Fig. 3).

Responses: we represent the responses of an operation as a subtree of the operation node. The sub-tree is collapsed by default indicating the presence of

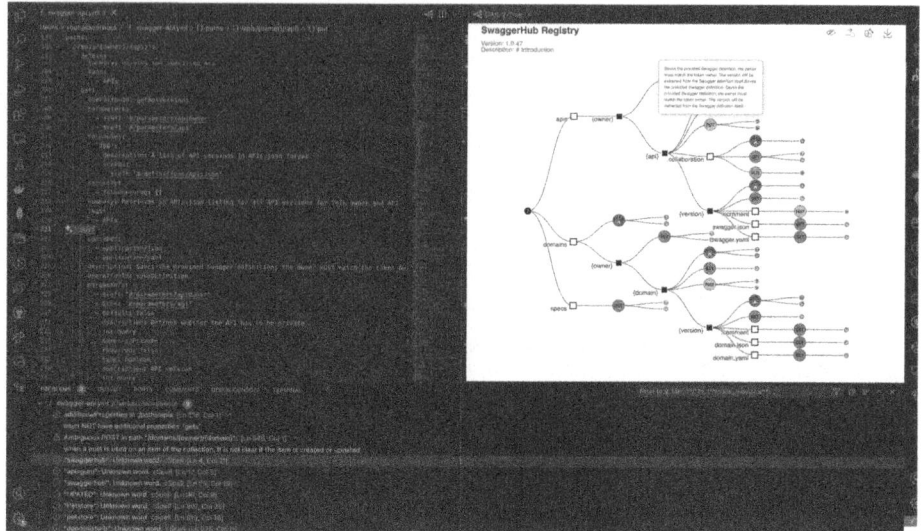

Fig. 3. Navigation between the diagram and the OAS code

responses. The user can expand it to see the details of the responses (Fig. 3). The root of the subtree is represented by a greater than icon ⊘ . Each response is represented by a node with the response status code. The response code is colored in green if it is a success code (2xx), in orange if it is a client-side error (4xx), and in red if it is a server-side error (5xx).

4 OAS2Tree Features

4.1 API Spec Validation Design Smells Detection

OAS2Tree can be used to detect smells in the API design described in the specification. The tool currently supports the detection design smells we empirically identified in our previous research study on API collection resource patterns [9]. The goal is to alert developers to potential design issues. The design smells detected by OAS2Tree include:

- **Ambiguous PUT and POST endpoints:** When an API contains both PUT and POST operations with similar paths, it can lead to confusion.
- **Create without delete:** When an API allows the creation of resources without providing a corresponding delete operation, it can result in data inconsistencies.
- **Delete without create:** When an API allows the deletion of resources without providing a corresponding create operation, it can lead to data loss.
- **Write-only endpoints:** When an API contains endpoints that only allow write operations without providing read capabilities, it can limit the usability of the API.

In addition, OAS2Tree validates the API specification against the OpenAPI Specification schema to ensure that it adheres to the standard, and highlights any errors or inconsistencies in the document and in the problems view as depicted in Fig. 3. For both smells and validation errors, the user can navigate to the problematic element in the OAS document by clicking on the error message.

4.2 Navigation of API Description Through the Tree Visualization

The tree structure visualization allows users to navigate through the API description easily. By expanding and collapsing nodes, users can explore the API structure and view the details of each endpoint, operation, and parameter. This interactive feature provides a comprehensive overview of the API architecture, enabling users to quickly locate specific endpoints and understand their functionalities. On mouse over, the editor highlights the corresponding element in the OpenAPI document in yellow. The description of the element is displayed in the tooltip. It is also possible to navigate from the problems view to the problematic element in the OpenAPI document by clicking on the error message.

4.3 Web Version of OAS2Tree

OAS2Tree is also available as a standalone web application, allowing users to visualize API structures without the need for a development environment. The web app version provides the same features as the VSCode extension, including the ability to save the current specification being visualized and share it through a unique URL with other users. The web app version is particularly useful for users who want to quickly visualize or sketch an API design from their Web browser (Fig. 4). An additional feature that the web version offers is the ability to navigate a collection of API specification examples, and visualize them in the tree structure format. This feature can be useful for users who want to explore different API designs and understand the common patterns and structures used in API development, and how some features are documented in OpenAPI by other API designers[1].

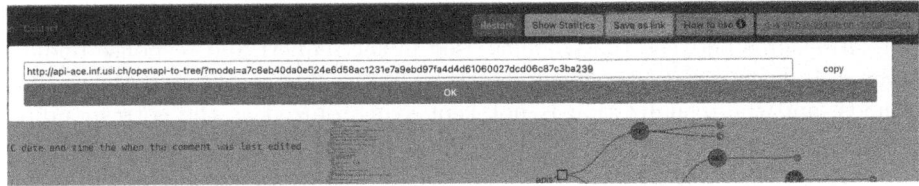

Fig. 4. Save as URL functionality in OAS2Tree Web App

---

[1] http://api-ace.inf.usi.ch/openapi-to-tree/navigate-apis?limit=50&page=13.

5 Usage Scenarios

OAS2Tree can be used by both API providers and API consumers in various use-case scenarios across the API development lifecycle.
As an API Designer & Developer:

- API Design and Documentation: During the initial stages of API design, OAS2Tree can be employed to visualize and refine the API structure. Since the visualization can also be generated from not fully complete OAS, it can serve as a tool for sketching partial API structures, and have an early overview of it.
- API Review and Design Validation: Designers can use OAS2Tree not only to ensure conformance to the OpenAPI standard, but also to ensure consistency in endpoint naming, parameter usage, and selection of HTTP methods.
- API Evolution: The tool can help find the adequate place of an extension.

As an API Consumer:

- Functionalities exploration: Client developers can use OAS2Tree to explore the endpoint structure and assess whether the API meets their requirements.
- Documentation navigation: OAS2Tree can be used to navigate the API documentation and quickly locate specific endpoints and operations, by hovering over the tree nodes, the corresponding element in the OpenAPI document is highlighted.
- Visual comparison: OAS2Tree can be used to visually compare the structure of different versions of an API, or the structures of different APIs.

6 Related Work

Most of the currently widely used API lifecycle management tools in the market[2][3][4], including paid ones, offer a Postman-like interface in their 'API Design environment', where the user can interact with the API, and test it. However, none of them offers a tree-like visualization of the API structure, that can be used to understand the overall API structure and navigate the documentation.

OAIE Sketcher[5] offers another way to visualize the endpoints by emphasizing the relationships between the schemas used in the request and response bodies. However, it lacks the hierarchical structure of the API paths and the HTTP methods, and the visualization is not kept synchronized as the corresponding textual specification changes.

The tool OpenAPItoUML [3] generates UML models from OpenAPI definitions, providing a means to visualize both API endpoint structures and API

---

[2] https://xapihub.io/features/designAndDev.
[3] https://stoplight.io/drive-api-results.
[4] https://apigit.com/why-apigit/api-design.
[5] https://raw.githack.com/OAIE/oaie-sketch/master/sketch.html.

data models using class diagrams. However, it does not provide a tree-like visualization of the API paths and operations, and it does not support the detection of any design smells in the API specification or data schema issues.

OAS2Tree differentiates itself from existing tools [8] by focusing specifically on rendering OpenAPI descriptions as tree-like visualizations.

7 Conclusion and Future Work

OAS2Tree is a REST API visualization tool, available as a Visual Studio Code extension and a web application, that transforms OpenAPI Specifications into a simple visual tree representation. It supports OAS v2.0, v3.0, and v3.1. It provides a side panel in the VS Code editor to display the API structure as a tree. The tool can be employed by API designers to visualize and refine the API structure during the initial design phase, ensuring consistency in endpoint naming, parameter usage, and HTTP method selection. It can also be used to validate the API design and detect potential design smells early on. API consumers can use OAS2Tree to explore the functionalities of an API and understand whether it meets their requirements. In addition, it contains a navigation feature that can help to quickly locate the textual description of specific endpoints and operations.

In future work, we plan to extend the tool to support detecting additional design smells and enhance the representation of the detected issues on the tree visualization to ease their location in the overall API structure. We are also experimenting with embedding the visualization as part of the API documentation generated from the OpenAPI description.

References

1. OAS2Tree. http://api-ace.inf.usi.ch/openapi-to-tree/
2. OAS2Tree. https://marketplace.visualstudio.com/items?itemName=oas2tree.oas2tree
3. Ed-douibi, H., Cánovas Izquierdo, J.L., Cabot, J.: OpenAPItoUML: a tool to generate UML models from OpenAPI definitions. In: Mikkonen, T., Klamma, R., Hernández, J. (eds.) ICWE 2018. LNCS, vol. 10845, pp. 487–491. Springer, Cham (2018). https://doi.org/10.1007/978-3-319-91662-0_41
4. Grent, H., Akimov, A., Aniche, M.: Automatically identifying parameter constraints in complex web APIs: a case study at adyen. In: 2021 IEEE/ACM 43rd International Conference on Software Engineering: Software Engineering in Practice (ICSE-SEIP), pp. 71–80. IEEE (2021)
5. OpenAPI Initiative: OpenAPI specification (2021). https://spec.openapis.org/oas/v3.1.0
6. Postman: Postman (2021). https://www.postman.com/
7. Serbout, S., Di Lauro, F., Pautasso, C.: Web APIs structures and data models analysis. In: 2022 IEEE 19th International Conference on Software Architecture Companion (ICSA-C), pp. 84–91. IEEE (2022)

8. Serbout, S., Hurtado, D.C.M., Pautasso, C.: Interactively exploring API changes and versioning consistency. In: 11th IEEE Working Conference on Software Visualization (VISSOFT 2023), Bogota, Colombia, pp. 28–39. IEEE (2023)
9. Serbout, S., Pautasso, C., Zdun, U., Zimmermann, O.: From OpenAPI fragments to API pattern primitives and design smells. In: 26th European Conference on Pattern Languages of Programs (EuroPLoP), pp. 1–35 (2021)
10. Zimmermann, O., Stocker, M., Lübke, D., Zdun, U., Pautasso, C.: Patterns for API Design: Simplifying Integration with Loosely Coupled Message Exchanges. Addison-Wesley Signature Series (Vernon). Pearson Education (2023)

A Multi-variant Benchmark for Microservice Systems in Software Engineering Research

Tomas Cerny[1](✉), Md Showkat Hossain Chy[1],
Muhmmad Ashfakur Rahman Arju[2], Korn Sooksatra[3],
Amr S. Abdelfattah[1], and Valentina Lenarduzzi[4]

[1] Systems and Industrial Engineering, University of Arizona, Tucson, AZ 85721, USA
tcerny@arizona.edu
[2] Computer Science, Montana State University, Bozeman, MT, USA
[3] Computer Science, Baylor University, Waco, TX, USA
[4] M3S, University of Oulu, Oulu, Finland

Abstract. Microservice architecture became the mainstream for cloud-native systems. While many microservice system benchmarks have been introduced to the scientific community, there is still a notable gap since the benchmarks do not offer architectural variants of the same system with identical functionality. For instance, research engaged mono-to-micro decomposition, but proposed methods use textbook or tutorial systems. Moreover, different microservice granularities in the system provide different architectural trade-offs. This paper extends an established microservice benchmark with two new variants, including a system monolith and a version with 20 microservices. Equivalent functionality is validated across the three benchmark variants.

Keywords: Microservices · Benchmarks · Mono-to-Micro · System Variants · Trade-offs · Software Architecture

1 Introduction

Software system benchmarks play a crucial role in software engineering as they enable a standardized way to evaluate scientific advancements and contributions. They allow developers and researchers to measure various metrics to assess the efficiency of their newly proposed methods against established baselines. Benchmarks help to accomplish fair comparisons across scientific works, objectively assess the strengths and weaknesses of various approaches, and make informed decisions about which solution best suits particular needs.

With hundreds of open-source microservice benchmarks in the community [6,10], only a few are frequently used in scientific literature. Yet, software architects should be critical and ask what led to specific modularization choices applied to these system benchmarks. For instance, one of the open challenges in microservice research [2] is monolith system decomposition into microservices, where legacy systems are turned into up-to-date mainstream trends. Many have approached this topic by using Machine Learning (ML) to drive such decomposition, but they lack proper assessment and benchmarks for validation [5]. In fact, different software architects could end up with different decomposition of microservices.

With the advancements in ML capabilities, we predict new trends will be applied to continuously improve the microservice system architecture, offering the right service granularity that fits the infrastructure/hardware constraints or the end user behavior. To foster such advancements, we observe the growing need for new types of system benchmarks, benchmarks that would provide system variants. For instance, a monolithic system version should co-exist with the microservice counterpart. Yet, different granularities of the same system could exist in terms of microservices to enable research on different trade-offs.

To have quick value for the community, an established benchmark should be extended to establish such variants, allowing incremental advancements. While some benchmarks [6] lead with the number of references, they are often rather small, with less than ten microservices. Microservices were designed for large systems. A mid-size benchmark over forty microservices might better serve the broader perspective of a more realistic industry-size system.

This paper extends the Train Ticket system benchmark [14] with two variants. One benchmark is based on monolithic architecture. The other considers the monolithic variant's dependency graph and aims to optimize modularity. It gives a variant of the system with 20 microservices.

2 On Microservice Benchmark Variants

We focused on creating two additional architecturally distinct versions of the baseline system: (1) a monolithic system variant and (2) a modularity-optimized microservice system variant. These variants are designed to offer the software engineering community valuable insights into the architectural nuances of transitioning between monolithic and microservice architectures, providing a fertile ground for exploring the implications of such transitions on system design, resource utilization on the cloud, selected scalability, performance, etc. Moreover, we detail the validation of these variants, aiming to ensure that the functionality across all variants remains the same.

Baseline Benchmark Selection: There is a limited amount of work aimed at the curation of datasets of microservice system benchmarks. The foundational work was introduced in 2018 by Brogi *et al.* [3], who produced a dataset containing 24 complex microservice systems with more than five microservices for demo purposes. Followed by Rahman *et al.* [10] in 2019, a curated dataset of 20 GitHub projects was introduced. Apart from this, other works compiled a microservice dataset to asses their works. Márquez and Astudillo [9], mentioned 30 projects to assess architectural patterns. Waseem et al. [13] investigated five projects on

GitHub, and Schneider et al. [11] proposed a dataset of 17 GitHub projects. Most recently, d'Aragona [6] presented a labeled dataset of 378 microservice projects.

When we consider benchmarks provided by d'Aragona [6], we can easily recognize that many benchmarks are small-sized. Some have industry support, but these are typically tools or instruments rather than enterprise system representatives. Microservices were invented to cope with growing system sizes, and thus, a suitable candidate should have at least 12+ microservices. Moreover, given industry practitioner feedback, most microservices take advantage of enterprise development frameworks like Spring or Enterprise Java. While this does not imply any kind of superiority of such a framework, it leads to our baseline benchmark selection. Across the research we were involved in, an open-source benchmark called Train Ticket has been mentioned, used, and referenced many times. Train Ticket was introduced at the ICSE conference [14], and the introduction work has 150 citations on Google Scholar.

The Baseline System Variant: With its robust microservice architecture, the baseline Train Ticket system was introduced in 2018 and advanced through multiple system versions. It stands as a pivotal benchmark in distributed systems design and implementation. We considered the benchmark version 1.0.0 given its stability. This version comprises 47 microservices, this system intricately simulates a comprehensive Train Ticket booking platform, encapsulating the complexities and advantages of microservices. Its technological diversity, incorporating Java, Node.js, Python, Go, and frameworks like Spring Boot and Django, reflects the multifaceted nature of real-world software ecosystems.

However, we focused on 42 of 47 microservices as the baseline for evaluating architectural variations. These 42 microservices encapsulate the primary business logic of the train ticket system and reflect the multifaceted nature of real-world software ecosystems. While part of the broader system architecture, the excluded services perform specific supportive tasks or enhancements such as user interface management and specialized data processing, peripheral or not in current use.

This architecture serves as a baseline for exploring the nuances of microservices and also establishes a standard for evaluating the impact of architectural variations on system functionality and performance.

The Monolithic System Variant: Transforming the baseline, bustling ecosystem of 42 Java-based microservices into a single, streamlined monolithic system was not just about lumping components together. It involved a thoughtful examination of how each piece of the system interacts, ensuring that we maintain the system's functionality while simplifying its architecture. We navigated through the complexities of integrating various services, solving puzzles related to how classes communicate, and making sure everything fits perfectly in a single, harmonious structure[1]. Validation is detailed in Sect. 3.

The Modularity-Driven System Variant: A comprehensive reengineering initiative was undertaken to meticulously design an architecture that adeptly

---

[1] We have made available the monolithic and microservice versions of the Train Ticket system benchmark in https://zenodo.org/records/11773560.

balanced service granularity. This endeavor was rooted in our steadfast commitment to augmenting scalability and maintainability while safeguarding the intricate functional aspects of the system.

Our modularity-driven approach was structured to address multifaceted objectives. Firstly, it aimed to foster robust interconnectedness among all classes within each microservice. Concurrently, efforts were directed towards curtailing excessive inter-microservice communication, with certain classes being permitted to transcend boundaries to facilitate this objective. Additionally, a paramount emphasis was placed on minimizing class duplication across microservices to uphold the tenets of code maintainability.

However, the simultaneous resolution of these three objectives posed a formidable computational challenge, given the NP-complete nature of the problem. To confront this inherent complexity, our approach harnessed a machine-learning-based strategy [5], thereby optimizing the problem and yielding a pragmatic solution. Ultimately, our methodological framework proposed a decomposed structure comprising 20 microservices, identified as the optimal equilibrium between modularity, system cohesion, and performance[1].

The implementation of this refined version effectively mitigated architectural concerns by strategically eliminating dependencies and shared resources previously dispersed across microservices-an approach acknowledged as counterproductive in cloud-native environments. While our algorithmic recommendations introduced an innovative decomposition, meticulous adherence to these guidelines resulted in the creation of a more modular yet functionally equivalent system. Validation is detailed in Sect. 3.

The transition towards this modularity-driven architecture engendered the preservation of core functionalities and a discernible enhancement in system performance metrics, characterized notably by reduced CPU and memory utilization. This version stands as a pivotal exemplar within the academic discourse on architectural best practices, underscoring the transformative potential of deliberate adjustments to service granularity in optimizing system maintainability and efficiency.

3 Validation

A robust validation was performed to ensure the fidelity and functional equivalence of the Train Ticket system variants. A repeatable test suite aimed to test system functionality provided through the user interface or the endpoints. Proceeding with the validation execution entailed evaluating the process outcomes and assessing the test statuses. Moreover, a thorough assessment of the generated dynamic traces was conducted to validate the benchmark and produce a comprehensive trace dataset demonstrating the runtime execution behavior.

Validation Test Suite: To validate the Train Ticket benchmark using a test suite, Smith et al. [12] developed a best-effort test suite for the Train Ticket benchmark, striving for complete coverage [1]. This validation process utilizes their test suite, Train Ticket Selenium Tests[2], which includes a broad range of meticulously designed End-to-End (E2E) test cases (51 use cases).

This test suite serves as a rigorous benchmark, enabling us to systematically assess and validate the operation of the monolithic and modularity-driven versions. Through this validation process, we establish a baseline for comparing architectural variants, ensuring that each new variant faithfully replicates the functionalities and services of the baseline system variant, thereby affirming their reliability as benchmarks for software engineering research.

Dynamic Traces Assessment: The Train Ticket system initially utilized Apache SkyWalking for dynamic tracing, capturing interactions limited primarily to API and service call levels. To enhance trace granularity, we aimed to extend tracing to the method level, ensuring that each method invocation within any API call generates a corresponding node within the trace graph. This approach facilitates a comprehensive trace graph where nodes are interconnected by a '`traceId`', delineating the flow of method calls across microservice boundaries. For instance, if a method in one microservice triggers an API call to another, the control flow in the recipient microservice is depicted as a descendant node in the trace graph of the originating service.

To achieve this enhanced tracing capability, we transitioned to OpenTelemetry [7]. We employed a specific OpenTelemetry plugin named '`methods`' to capture method-level details without altering existing code. This plugin dynamically attaches trace information to each method invocation as configured, leveraging static analysis to automatically enumerate all method declarations in the system, including those dynamically generated at runtime, such as methods introduced by the Lombok library. The traces are shared in the Zenodo package[1] alongside the benchmark.

Jaeger [8] was utilized as a tracing and logging tool in our system variants because it offers detailed insights into how requests are processed across the various services. By using Jaeger, we benefit from its ability to identify bottlenecks and performance issues within our system, which allows for more efficient troubleshooting and optimization of service interactions. The output from Jaeger includes visual representations like trace timelines, which show the flow of requests through our services, and dependency diagrams that illustrate how services interact with each other. This comprehensive visibility helps us ensure that our system operates smoothly and efficiently.

---

[2] Train Ticket Test Benchmark: https://github.com/cloudhubs/microservice-tests.

4 Discussions

In transforming the baseline 42-microservice system, which predominantly used Java-based Spring Boot services, into a monolithic system and then a modular 20-microservice system, we aimed to address modularity and key architectural concerns. Our modularity-driven transition strategy ensured that while the system was restructured, the functional equivalence was maintained, evidenced by the successful passage of all tests from an established testing benchmark [12]. The source code for these new system variants is provided as a Zenodo package[1], with supplementary materials demonstrating functional consistency.

In the baseline variant, we observed an anti-pattern of business and other class logic injections, which led to tightly coupled components that hindered the maintainability of self-contained microservices [4]. This challenge was one of the primary motivations for exploring other architectural variants. The monolithic system variant consolidated functionalities that were previously scattered across multiple services, simplifying deployment and operational processes. However, this consolidation brought its own set of challenges, particularly in scalability and flexibility. The monolithic architecture, while simpler to manage, proved rigid in accommodating changes, which could impact the entire system.

The transition to the 20-microservice variant is detailed through Table 1. It shows classes from the baseline variant microservices divided into the 20-microservice variant. This was a well-engineered approach targeting modularity. That is why some baseline microservice classes (i.e., ts-order-other-service) divide into multiple microservices of the new variant, and some become partially duplicated across selected microservices, which is not a bad design practice but a consequence of node duplication strategy in the statistical model.

The transition demonstrated a shift from the baseline design by increasing the total lines of code and number of classes, which is due to a bad design choice in the baseline variant where the design used the parent-child project structure strategy and the allowance for class replication. The parent-child strategy in the microservice paradigm is a poor choice as every time change occurs, the whole system needs to be redeployed again. Due to this limitation in baseline, we had to duplicate common config-related classes in multiple microservices. The SourceMonitor[3] analysis shows that while the 20-microservice system had a higher codebase and class count, it benefited from reduced cyclomatic complexity and better separation of concerns.

---

[3] SourceMonitor: https://www.derpaul.net/SourceMonitor.

Table 1. Mapping microservice classes between the baseline and 20-microservice system variants

	ts-admin-service	ts-assurance-service	ts-auth-service	ts-cancel-service	ts-config-service	ts-consign-service	ts-contacts-service	ts-delivery-service	ts-food-service	ts-notification-service	ts-order-related-service	ts-order-service	ts-preserve-service	ts-price-service	ts-rebook-service	ts-route-service	ts-security-service	ts-station-service	ts-travel-service	ts-user-service
ts-admin-basic-info-service							✓											✓		
ts-admin-order-service	✓										✓									
ts-admin-route-service		✓		✓												✓				
ts-admin-travel-service		✓											✓							
ts-admin-user-service		✓		✓	✓			✓		✓					✓					✓
ts-assurance-service		✓	✓							✓	✓	✓	✓		✓					✓
ts-auth-service	✓	✓	✓	✓	✓	✓		✓		✓	✓	✓			✓					✓
ts-basic-service		✓								✓										
ts-cancel-service				✓	✓										✓	✓				✓
ts-common	✓	✓	✓	✓	✓	✓	✓	✓	✓	✓	✓	✓	✓	✓	✓	✓	✓	✓	✓	✓
ts-config-service	✓	✓		✓	✓	✓				✓	✓				✓	✓		✓	✓	
ts-consign-price-service		✓		✓	✓					✓						✓				
ts-consign-service		✓		✓	✓	✓				✓	✓	✓			✓					✓
ts-contacts-service		✓		✓			✓				✓	✓		✓				✓	✓	✓
ts-delivery-service	✓			✓				✓	✓		✓	✓	✓		✓					✓
ts-execute-service	✓			✓	✓										✓					✓
ts-food-delivery-service								✓	✓											
ts-food-service	✓			✓			✓		✓	✓		✓	✓	✓	✓					✓
ts-inside-payment-service		✓	✓	✓	✓	✓		✓			✓				✓	✓				✓
ts-notification-service	✓			✓	✓	✓			✓	✓	✓	✓			✓					✓
ts-order-other-service	✓	✓		✓	✓	✓	✓	✓	✓	✓	✓	✓	✓	✓	✓	✓	✓	✓	✓	✓
ts-order-service	✓	✓		✓	✓	✓	✓	✓	✓	✓	✓	✓	✓	✓	✓	✓	✓	✓	✓	✓
ts-payment-service		✓		✓	✓	✓		✓			✓				✓	✓				✓
ts-preserve-other-service	✓			✓				✓		✓	✓		✓	✓						✓
ts-preserve-service	✓			✓				✓		✓	✓		✓	✓						✓
ts-price-service		✓			✓	✓	✓				✓	✓				✓				
ts-rebook-service												✓		✓					✓	✓
ts-route-plan-service												✓		✓						✓
ts-route-service		✓		✓			✓			✓	✓	✓	✓	✓	✓	✓		✓	✓	✓
ts-seat-service												✓							✓	✓
ts-security-service			✓	✓	✓					✓					✓	✓				
ts-station-food-service			✓	✓	✓		✓	✓	✓											
ts-station-service		✓		✓		✓			✓		✓	✓			✓	✓			✓	✓
ts-train-food-service		✓	✓			✓	✓	✓										✓		
ts-train-service		✓	✓			✓				✓	✓	✓	✓	✓				✓	✓	✓
ts-travel-plan-service	✓										✓	✓	✓	✓					✓	
ts-travel-service	✓	✓		✓			✓		✓	✓	✓	✓			✓		✓		✓	✓
ts-travel2-service	✓	✓		✓			✓		✓	✓	✓	✓			✓		✓		✓	✓
ts-user-service	✓	✓	✓	✓	✓	✓		✓		✓	✓				✓				✓	
ts-verification-code-service			✓		✓	✓									✓					
ts-wait-order-service	✓	✓													✓	✓			✓	✓

5 Conclusions

System granularity is a key factor that can impact its architectural stability and quality attributes. This paper contributes to facing the microservice system benchmark shortage by introducing a multi-variant benchmark derived from a well-established open-source Train Ticket system. It introduces a monolithic variant along with two microservice system variants of different granularities, consisting of 42 and 20 microservices, respectively. These variants were validated using an open-source E2E test suite, demonstrating equivalent functionality. Furthermore, the validation execution process yielded dynamic traces, which are refined and provided as a dataset ready for research investigation that can nurture new case studies to explore the execution behavior of a monolithic system and two microservice benchmarks of different granularities.

Future work will extend the variability by implementing selected Train Ticket microservices in different platforms to enable polyglot analysis and use serverless for selected system features to provide more trade-off options to the community.

References

1. Abdelfattah, A.S., Cerny, T., Yero, J., Song, E., Taibi, D.: Test coverage in microservice systems: an automated approach to E2E and API test coverage metrics. Electronics **13**(10), 1913 (2024)
2. Bogner, J., Fritzsch, J., Wagner, S., Zimmermann, A.: Industry practices and challenges for the evolvability assurance of microservices. Empir. Softw. Eng. **26**(5), 104 (2021). https://doi.org/10.1007/s10664-021-09999-9
3. Brogi, A., Canciani, A., Neri, D., Rinaldi, L., Soldani, J.: Towards a reference dataset of microservice-based applications. In: Cerone, A., Roveri, M. (eds.) SEFM 2017. LNCS, vol. 10729, pp. 219–229. Springer, Cham (2018). https://doi.org/10.1007/978-3-319-74781-1_16
4. Cerny, T., Abdelfattah, A.S., Maruf, A.A., Janes, A., Taibi, D.: Catalog and detection techniques of microservice anti-patterns and bad smells: a tertiary study. J. Syst. Softw. **206**, 111829 (2023). https://doi.org/10.1016/j.jss.2023.111829. https://www.sciencedirect.com/science/article/pii/S0164121223002248
5. Chy, M.S.H., Sooksatra, K., Yero, J., Cerny, T.: Benchmarking micro2micro transformation: an approach with GNN and VAE. Clust. Comput. (2024). https://doi.org/10.1007/s10586-024-04526-z
6. d'Aragona, A.D., et al.: A dataset of microservices-based open-source projects. In: 2024 IEEE/ACM 21st International Conference on Mining Software Repositories (MSR), pp. 215–219 (2024)
7. Cloud Native Computing Foundation: OpenTelemetry Project Journey Report | CNCF (2023). https://www.cncf.io/reports/opentelemetry-project-journey-report/
8. Jaeger Authors: Jaeger: open source, distributed tracing platform (2024). https://www.jaegertracing.io/
9. Márquez, G., Astudillo, H.: Actual use of architectural patterns in microservices-based open source projects. In: 2018 25th Asia-Pacific Software Engineering Conference (APSEC), pp. 31–40 (2018)

10. Rahman, M.I., Taibi, D.: A curated dataset of microservices-based systems. In: Joint Proceedings of the Summer School on Software Maintenance and Evolution. CEUR-WS (2019)
11. Schneider, S., Özen, T., Chen, M., Scandariato, R.: microsecend: a dataset of security-enriched dataflow diagrams for microservice applications. In: 2023 IEEE/ACM 20th International Conference on Mining Software Repositories (MSR), pp. 125–129. IEEE (2023)
12. Smith, S., et al.: Benchmarks for end-to-end microservices testing. In: 2023 IEEE International Conference on Service-Oriented System Engineering (SOSE), Los Alamitos, CA, USA, pp. 60–66. IEEE Computer Society (2023). https://doi.org/10.1109/SOSE58276.2023.00013
13. Waseem, M., Liang, P., Shahin, M., Ahmad, A., Nassab, A.R.: On the nature of issues in five open source microservices systems: an empirical study. In: 25th International Conference on Evaluation and Assessment in Software Engineering, EASE 2021, pp. 201–210 (2021)
14. Zhou, X., Peng, X., Xie, T., Sun, J., Xu, C., Ji, C., Zhao, W.: Benchmarking microservice systems for software engineering research. In: Proceedings of the 40th International Conference on Software Engineering: Companion Proceedings, ICSE 2018, pp. 323–324. ACM, New York (2018)

Semantics Enhancing Model Transformation for Automated Constraint Validation of Palladio Software Architecture to MontiArc Models

Sebastian Weber[1]($^\boxtimes$), Jörg Henß[1], Bahareh Taghavi[2], Thomas Weber[2], Sebastian Stüber[3], Adrian Marin[3], Bernhard Rumpe[3], and Robert Heinrich[2]

[1] FZI Research Center for Information Technology, Karlsruhe, Germany
{sebastian.weber,henss}@fzi.de
[2] Karlsruhe Institute of Technology, Karlsruhe, Germany
{bahareh.taghavi,thomas.weber,robert.heinrich}@kit.edu
[3] Software Engineering, RWTH Aachen University, Aachen, Germany
{stueber,marin,rumpe}@se-rwth.de

Abstract. Component-based software architecture allows software architects to design systems by composing components with syntactically defined interfaces. These models can be used for the analysis and prediction of the functional and non-functional properties of the system. While tools for the modeling and analysis of such systems, e.g., the Palladio approach, support the syntactic validation of the composition, they lack the capability to validate the semantic composition. If, e.g., one component requires and one provides an integer value, they can be composed, independently of whether this composition is actually semantically sound. To support software architects in the semantic validation of their system models, we propose a model transformation tool, that allows to transform system models from Palladio models to MontiArc models, enrich them with semantic constraints and validate these constraints with the MontiArc workbench. We present exemplary results of this transformation and validation applied to a simplified model of a component-based simulator of the Palladio approach.

Keywords: Semantic Constraint Validation · Software Architecture · Model Transformation · Palladio · MontiArc

1 Introduction

Software plays an ever more important role in society and economy. To ensure it fulfils the requested functionality and quality properties, such as performance and security, a multitude of different analysis techniques are available. Because these analysis techniques are designed to evaluate specific questions, they use different modeling formalisms tailored to these questions. The Palladio simulators

[15] and the Palladio Component Model (PCM) [15] are examples and while they enable the architectural modeling and analysis of component-based software systems, they do not support the validation of the composition of such components beyond a syntactic level, i.e., as discussed in [6]. As long as the syntactic interfaces between the components match, they can be composed, but it is crucial to also verify that all the components can work together and communicate properly by semantic validation. Because Palladio aims at analyzing quality properties of systems, we do not introduce support for semantic constraints directly in Palladio, but instead we introduce a tool that allows for the transformation from a PCM model to a MontiArc [4] model. MontiArc supports the textual modeling of component&connector systems and allows the validation of semantic constraint specified at the ports of the components, which are connected through the connectors. These constraints are specified in an additional model to avoid adding more complexity to the PCM and enrich the MontiArc model generated from the PCM. Our tool is publicly available at Github (https://github.com/FeCoMASS/Model-Transformation-for-Automated-Constraint-Validation) and an explanatory video is available at https://fecomass.github.io/fecomass/videos/.

Contributions: Our first contribution is the transformation of a PCM model to a MontiArc. The second contribution is the incorporation of constraints within a MontiArc model and the last contribution is the addition of constraint checking to MontiArc. Figure 1 shows the transformation and enrichment process our tool applies to check the semantic validity of the composition of the system modeled in the PCM.

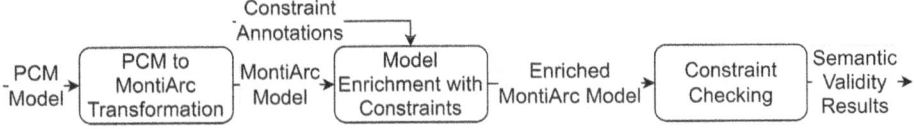

Fig. 1. Transformation Process and Artifacts

The following Sect. 2 introduces the Palladio approach, its simulators and the modeling formalism PCM. In addition, the model of a simplified exemplary system is presented. Section 3 introduces MontiArc and shows how the exemplary system is modeled there. The next Sect. 4 explains the automatic transformation between PCM and MontiArc, followed by Sect. 5 introducing how the constraints in the MontiArc model are checked to validate the modeled system. Section 6 concludes the paper with a discussion about the current state of the tool and an outlook on future work planned to expand it.

2   Palladio and Running Example

Palladio [15] is an approach for simulating software architecture, aiming to analyze and predict performance, among other quality properties. The tooling that implements the Palladio approach is known as Palladio-Bench [7]. The Palladio Component Model (PCM) is a domain-specific modeling language and is composed of multiple sub-models, each targeting a particular developer role. Component developers contribute by specifying behavioral aspects of their components and interfaces in the repository model. Subsequently, system architects leverage these repositories to assemble concrete component-based software systems in the assembly model. Meanwhile, system deployers focus on modeling the resource environment and allocating components across different resources. Business domain experts are also responsible for providing usage models that describe critical usage scenarios and outline user behavior. Palladio facilitates model evaluation through simulation, enabling the prediction of performance metrics like response times and hardware utilization under specified workloads.

Running Example: We use a simplified model of Slingshot [9], the latest simulator for Palladio based on an event-driven architecture, to showcase the PCM models our tool requires as depicted in Fig. 2. Slingshot currently comprises three simulation components: the *UsageSimulation*, the *SystemSimulation*, and the *ResourceSimulation* component. In order to start a simulation, the *UsageSimulation* component looks up the workload from a usage scenario and begins interpretation based on its parameters. The connection between the *UsageSimulation* and *SystemSimulation* components occurs via user requests, which model service calls within the system. In addition, the *SystemSimulation* component and the *ResourceSimulation* component are connected by requesting resource demands.

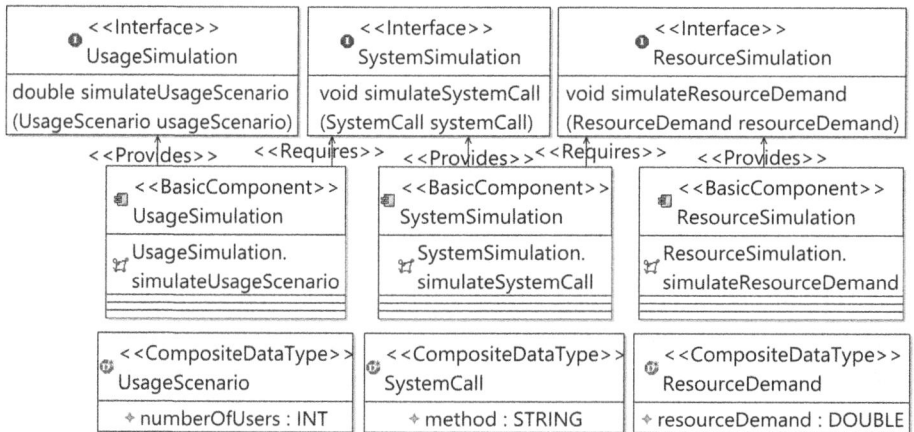

Fig. 2. Repository Model of Slingshot

The components of the repository are instantiated in the assembly model shown in Fig. 3. These assembly contexts are connected through directed connectors to either delegate from a system role to an assembly context role or between provided and required assembly context roles.

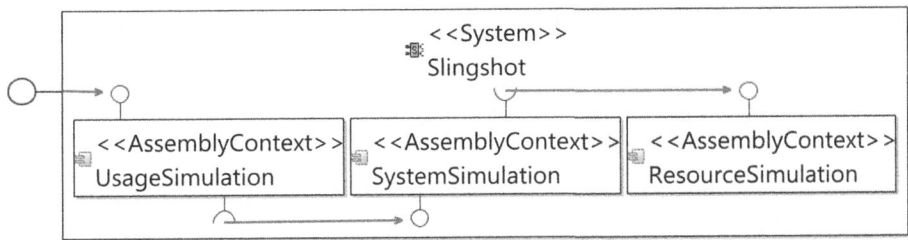

Fig. 3. Assembly Model of Slingshot

3 MontiArc

MontiArc (https://github.com/MontiCore/montiarc) [4] is a textual modeling language to describe component&connector systems. The components receive input messages and sent output messages via typed and directed ports. Only over these explicitly defined ports, communications is possible. This reduces hidden links.

MontiArc has a precisely defined semantic foundation in FOCUS [2]. The textual MontiArc models can be mapped into the mathematical FOCUS space to understand the semantics of the models [5]. This enables formal verification of MontiArc models at design time. The formal proofs are performed using MontiBelle [11,12] and the interactive theorem prover Isabelle [14]. In [10] this approach is demonstrated in cooperation with Airbus to verify properties over an uplink feed of an avionic system.

To leverage of MontiArc's simulation and verification capabilities we model Slingshot's component repository model as seen in Fig. 2 in MontiArc. Listing 1.1 shows an excerpt of the generated MontiArc model with constraints, restricting the number of users in a usage scenario to a positive number. Slingshot is modeled as the system component which is further decomposed into atomic subcomponents representing simulation components. Each sub-component instantiates their respective component definitions, which in turn define their ports in accordance to the event-driven data-flow paradigm employed in Slingshot/Palladio. The sub-components are instantiated and communicate through connections of their ports.

```
component Slingshot {
  port <<condition = "x.numberOfUsers > 0
                  && x.numberOfUsers < 100">>
       in UsageScenario usageScenario;
  component UsageSimulation {
    port <<condition = "x.numberOfUsers > 0">>
         in UsageScenario usageScenario;
  }
  UsageSimulation usageSimulation;
  usageScenario -> usageSimulation.usageScenario;
}
```

Listing 1.1. Simplified Excerpt of the MontiArc Model with Constraints

4 Transforming PCM to MontiArc

Our transformation uses the repository and assembly model of the PCM as inputs. The first step of the transformation is the extraction of complex data types from the PCM repository model. Primitive data types like `int` or `char` can be translated directly while complex data types require their own definition in a MontiArc class diagram model. In addition to complex data types defined in the PCM model, method signatures with two or more parameters are also transformed to MontiArc data types, to be able to specify constraints for them. The result of this step is the MontiArc class diagram containing all necessary data types for the description of the PCM system model as a MontiArc model.

In the next step, the PCM system model is transformed. First, the system, which is the root element of an assembly model, is transformed to the core component of the MontiArc model. Afterwards, the assembly contexts, which are instances of components specified in the repository, are transformed to sub components. The MontiArc component a PCM assembly context is transformed to have MontiArc ports according to the PCM roles of the PCM component this PCM assembly context encapsulates. The last part of this step is the transformation of PCM connectors, which connect PCM assembly contexts according to the roles from the encapsulated component, to MontiArc connectors. As last step, the generated MontiArc model is enriched with the constraints specified as Ecore annotations in the additional input model. Each annotation refers to a directed connector in the PCM assembly model and has two key value pairs, one describing the guarantee of the component the connector comes from and the other one describing the assumption of the component the connector goes to. These constraints are then mapped onto the corresponding MontiArc ports.

5 Constraint Checking

We present a solution to automatically provide consistency analysis to the source PCM models by checking the constraints produced during the transformation to

```
1 (declare-sort UsageScenario)
2 (declare-fun numberOfUsers (UsageScenario) (Int))
3 (declare-fun x () UsageScenario)
4 (assert (=> (and (< (numberOfUsers x) 100)
5                  (> (numberOfUsers x) 0))
6             (> (numberOfUsers x) 0)))
```

Listing 1.2. Translated Constraints in SMT-Lib

MontiArc. Constraints are embedded into stereotypes over ports of C&C architectural components. These employ a reduced form of Assumption/Guarantee formalism [2] implemented through an assertion paradigm similar to the programming language Eiffel [13]. The restriction in question is that the constraints are not expressed through formulas handling potentially infinite port flow histories but only instant values present on ports. This is first done by parsing the generated model and processing all constraint pairs defined with the *condition* stereotype. Processing constraints starts by parsing the constraints into the MontiCore expression framework [8] and performing a translation from these to SMT-Lib [1] formulas. Our implementation reuses an OCL to SMT translator restricted on the set of MontiCore expressions. The variant of the OCL accepted by the translator is described in [16]. Our restriction of the OCL language is introduced to facilitate a simple and decidable set of constraints that can be reliably solved by conventional SMT solvers, in addition to only handling quantifier-free constraints. Furthermore, complex objects transmitted on ports also require their datatypes to be represented and translated to SMT. Our approach uses implementation focused class diagrams and objects diagrams as presented in [16]. Microsoft's Z3 SMT solver [3] checks the SMT-Lib formulas. To formulate checkable constraints we build implications over the assumptions and the guarantee obligations. As port datatypes are instances of classes we introduce a SMT sort for each class. Listing 1.2 introduces the declaration of the *UsageScenario* sort and corresponding function *numberOfUsers* introduced for its attribute its attribute, all in accordance to the class diagram generated for the PCM model in Fig. 2. The return type of the function is represented by the SMT-Lib built-in Int sort.

The translation of the constraints takes place by introducing a variable x and and declaring it in the respective sort of the port's type, i.e., the sort generated for the class, here *UsageScenario*. The implications and logical operators used in constraints are then 1-to-1 mapped to SMT-Lib. The result of the checker is then constructed with a classic approach, i.e., to check validity we negate the implication and check for unsatisfiability. The result is then produced to the standard output with a statement about the validity of the implication and the the attributed ports. Listing 1.3 presents an output of the constraint checker for the valid Slingshot model, referencing the connected ports in question. If the negated implication is satisfiable, then the implication is not valid. Listing 1.4 presents the output of the constraint checker if we modify the target constraint

```
Constraint of Port usageScenario in Component Slingshot
   guarantees constraint of Port usageScenario in
   Component UsageSimulation
```

Listing 1.3. Output of the Constraint Checker for the Constraint-Enriched Model

```
[ERROR] Found error in port-constraint. Constraint of Port
    usageScenario in Component UsageSimulation does not
    follow from constraint of usageScenario in Component
    Slingshot.
Counterexample: objectdiagram x {
  usageScenario_0:UsageScenario {
    int numberOfUsers=1;
  };
}
```

Listing 1.4. Output of the Constraint Checker for a Negative Result in the Model

in listing 1.1 to require a positive number of users. The tool errors but not before processing all implications. The tool presents a counterexample in the form of an object diagram, performing a retranslation of the SMT-Lib model to the object-oriented representation.

6 Conclusion

We presented a tool for the automatic translation of software architecture models from the Palladio approach and the PCM to the architecture description language of MontiArc to support the validation of semantic constraints specified as annotations of model elements from the PCM. We believe this tool benefits the software architecture community by bridging the gap between both the PCM and MontiArc modeling and analysis approaches to support thorough system analysis without the need for repeated modeling of the same system in different formalisms. While the basic concepts shown in this paper can already be transformed from PCM to MontiArc, our tool currently does not support the full expressiveness of concepts that could be transformed. Furthermore, our tool does not support a retranslation of the result from MontiArc into the PCM. We currently only support a single method per interface, because we specify constraints only referencing connectors corresponding to an interface and not a method. In addition to supporting this, we plan to expand our tool to not only allow for the transformation and validation of PCM models but also the PCM metamodel and the simulator presented as an example in this paper. This allows for the validation of composition on the three different levels of model instance, metamodel, and analysis based on these models.

Acknowledgments. This work was funded by the DFG (German Research Foundation) – project number 499241390 (FeCoMASS), supported by the Collaborative Research Center "Convide" - SFB 1608 - 501798263 and supported by funding from the topic Engineering Secure Systems, KASTEL Security Research Labs funded by the Helmholtz Association (HGF).

References

1. Barrett, C., Stump, A., Tinelli, C., et al.: The SMT-lib standard: version 2.0. In: Proceedings of the 8th International Workshop on Satisfiability Modulo Theories (Edinburgh, UK), vol. 13, p. 14 (2010)
2. Broy, M., Stølen, K.: Specification and Development of Interactive Systems. Focus on Streams, Interfaces and Refinement. Springer, Heidelberg (2001). https://doi.org/10.1007/978-1-4613-0091-5
3. de Moura, L., Bjørner, N.: Z3: an efficient SMT solver. In: Ramakrishnan, C.R., Rehof, J. (eds.) TACAS 2008. LNCS, vol. 4963, pp. 337–340. Springer, Heidelberg (2008). https://doi.org/10.1007/978-3-540-78800-3_24
4. Haber, A.: MontiArc - Architectural Modeling and Simulation of Interactive Distributed Systems. Aachener Informatik-Berichte, Software Engineering, Band 24, Shaker Verlag (2016)
5. Harel, D., Rumpe, B.: Meaningful modeling: what's the semantics of "semantics"? IEEE Comput. J. **37**(10), 64–72 (2004)
6. Heinrich, R., Strittmatter, M., Reussner, R.H.: A layered reference architecture for metamodels to tailor quality modeling and analysis. IEEE Trans. Software Eng. **47**(4), 775–800 (2021)
7. Heinrich, R., et al.: The palladio-bench for modeling and simulating software architectures. In: Proceedings of the 40th International Conference on Software Engineering: Companion Proceedings, pp. 37–40 (2018)
8. Hölldobler, K., Kautz, O., Rumpe, B.: MontiCore Language Workbench and Library Handbook: Edition 2021. Aachener Informatik-Berichte, Software Engineering, Band 48, Shaker Verlag (2021)
9. Katić, J., Klinaku, F., Becker, S.: The slingshot simulator: an extensible event-driven PCM simulator (poster) (2021)
10. Kausch, H., Pfeiffer, M., Raco, D., Rath, A., Rumpe, B., Schweiger, A.: A theory for event-driven specifications using focus and MontiArc on the example of a data link uplink feed system. In: Groher, I., Vogel, T. (eds.) Software Engineering 2023 Workshops, pp. 169–188. Gesellschaft für Informatik e.V. (2023)
11. Kausch, H., Pfeiffer, M., Raco, D., Rumpe, B.: An approach for logic-based knowledge representation and automated reasoning over underspecification and refinement in safety-critical cyber-physical systems. In: Hebig, R., Heinrich, R. (eds.) Combined Proceedings of the Workshops at Software Engineering 2020, vol. 2581. CEUR Workshop Proceedings (2020)
12. Kausch, H., Pfeiffer, M., Raco, D., Rumpe, B.: MontiBelle - toolbox for a model-based development and verification of distributed critical systems for compliance with functional safety. In: AIAA Scitech 2020 Forum. American Institute of Aeronautics and Astronautics (2020)
13. Meyer, B.: Lessons from the design of the eiffel libraries. Commun. ACM **33**(9), 68–88 (1990)

14. Nipkow, T., Paulson, L.C., Wenzel, M.: Isabelle/HOL: A Proof Assistant for Higher-Order Logic. Lecture Notes in Artificial Intelligence, vol. 2283. Springer, Heidelberg (2002). https://doi.org/10.1007/3-540-45949-9
15. Reussner, R.H., Becker, S., Happe, J., Heinrich, R., Koziolek, A.: Modeling and Simulating Software Architectures: The Palladio Approach. MIT Press, Cambridge (2016)
16. Rumpe, B.: Modeling with UML: Language, Concepts, Methods. Springer, Cham (2016). https://doi.org/10.1007/978-3-319-33933-7

Extending a Low-Code Tool with Multi-cloud Deployment Capabilities

Fitash Ul Haq[1], Iván Alfonso[1(✉)], Armen Sulejmani[1], and Jordi Cabot[1,2]

[1] Luxembourg Institute of Science and Technology, Esch-sur-Alzette, Luxembourg
{fitash.ulhaq,ivan.alfonso,armen.sulejmani}@list.lu, jordi.cabot@uni.lu
[2] University of Luxembourg, Esch-sur-Alzette, Luxembourg

Abstract. Low-code emerged as an evolution of model-driven engineering to accelerate software delivery, and it continues to gain traction today. However, low-code tools and solutions have primarily focused on development, often neglecting or offering minimal support for the application deployment process, such as lacking capabilities for multi-cloud deployments. In this paper, we propose an extension of BESSER, an open-source low-code platform, to address the packaging and deployment of applications in multi-cloud environments. This extension includes the definition of a language and a grammar to enable the modeling of the deployment architecture, also enabling the specification of public and on-premises clusters. Additionally, we have developed code generators to automate the application packaging, and cloud provisioning and deployment using Terraform. The complete infrastructure is available in an open-source repository.

Keywords: low-code · deployment architecture · multi-cloud

Demo Video: http://tiny.cc/demo-video

1 Introduction

In recent years, low-code tools have been expanding both in academia and software industry. These tools enable developers to focus more effort on business-related tasks as development complexity is reduced. Low-code tools aim to reduce the amount of manual-coding required to accelerate software delivery by raising the level of abstraction to facilitate domain specification and ignore irrelevant technical details. It can be seen as an evolution or continuation of model-based approaches [4], whose foundations are based on model-driven engineering (MDE). However, low-code should not only focus on accelerating application development, but also on supporting different processes throughout the software lifecycle such as testing, deployment, monitoring, and adaptation.

Usually, low-code tools omit, or provide little support for modeling other aspects such as deployment architecture (also known as hardware architecture

---

F. Ul Haq and I. Alfonso—These authors contributed equally.

[6]), i.e., aspects related to infrastructure, clustering, virtualization technology, and app deployment. For example, the specification of multi-cloud deployments that use several cloud providers to deploy the different application components or hybrid deployments that include on-premises clusters, are usually not supported.

To address this concern, we have developed an extension of the BESSER low-code platform [2] to (1) enable the specification of the deployment architecture of the target software and (2) automate the packaging, provisioning and deployment of the system according to that deployment model. This extension aims to transform BESSER into a low-code platform that can also cover these final phases of the software development lifecycle.

The remainder of the paper is organized as follows: Sect. 2 presents an overview of the low-code tool extension. Details on modeling and code generation are introduced in Sect. 3 and 4 respectively. Related work is discussed in Sect. 5, and Sect. 6 concludes the paper.

2 Overview of Extended BESSER Platform

As part of this work, we have extended the BESSER platform (1) by equipping it with a complete backend generator capable of producing the schema of a database and a REST API providing an abstraction layer for CRUD operations that can be packaged as a docker container, and (2) by providing BESSER with the capability of modeling the software architecture and, based on that model, deploying applications on a cloud infrastructure, making it available for end users to use. To perform the latter, we have introduced a new metamodel and grammar to parse the deployment model in textual form (details in Sect. 3), and a code generator (see Sect. 4) to generate an Infrastructure as Code definition that can be directly run to provision and deploy applications on the cloud.

This architectural view is combined with the more "traditional" models defining the structural and behavioural aspects of the application. Indeed, at the core of BESSER we have B-UML (BESSER's Universal Modeling Language), a foundational language designed for specifying and modeling various aspects of a system. This language supports the creation of different types of models, including structural or data models, object models, graphical user interface models, and even specifying OCL constraints. B-UML models serve as the input for code generators that produce application code. BESSER provides code generators for various technologies such as SQLAlchemy, Django, and Python.

Figure 1 shows the overview of extended BESSER platform combining the two perspectives. Both types of models are defined using a textual notation that conforms to their respective grammars. The BESSER platform parses these models and creates their respective B-UML models. Code generators interpret these B-UML models to produce various software artifacts, including the scripts for configuring and deploying the application on the multi-cloud environment, which is accessible by users.

Internally, BESSER first parses the structural model and generates a B-UML structural model that conforms to B-UML metamodel. This B-UML metamodel is then used by code generators to produce the core code for the application

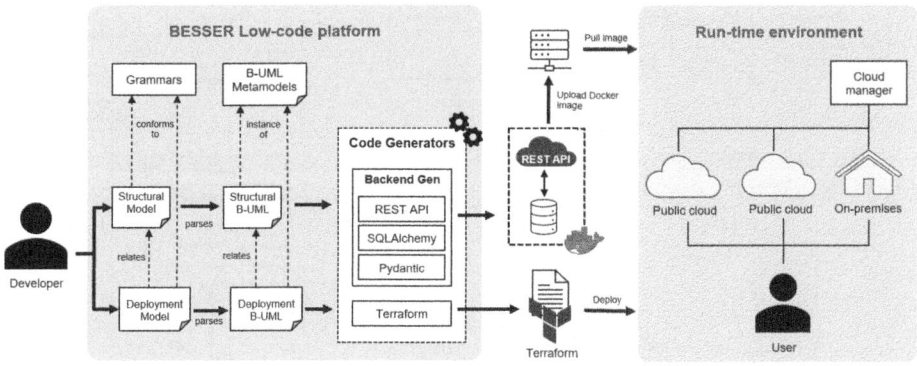

Fig. 1. Overview of extended BESSER platform

along with the database support. In a second step, the platform generates python scripts to containerize the application along with the database. BESSER uses this script to upload the image to the hub automatically. BESSER currently supports DockerHub for this purpose but can be easily extended for other repositories. In a third step, the platform parses the deployment model and generates a deployment B-UML model, that conforms to B-UML deployment architecture metamodel, which is then used by code generators to generate the Terraform[1] scripts. Terraform is a state-of-the-art open source tool developed to deploy applications on the cloud infrastructure. Finally, these Terraform scripts are run to configure the cloud infrastructure (including nodes, clusters, networks, etc.) and deploy the containerized application using the image from the hub (created in step 2). Additionally, a load balancer is also provisioned to distribute the web traffic between the nodes in the clusters.

3 A Modeling Extension for Specifying a Deployment Architecture

To deploy application on the cloud, BESSER platform requires two models from the user: (1) a structural model that contains information about the application itself and (2) a Deployment model that contains information about the target cloud-based deployment architecture.

Due to the space limitations, we do not discuss the structural model in this paper. More details on this model can be found in this previous work [2]. Therefore, the rest of this section focuses on the deployment model.

Figure 2 shows the metamodel we have introduced for modeling deployment architecture of cloud infrastructure. The metamodel contains a root *Deployment-Model* metaclass, which is the main construct of the model. It comprises all the information about the cloud infrastructure. A *DeploymentModel* can have one or

---

[1] https://www.terraform.io/.

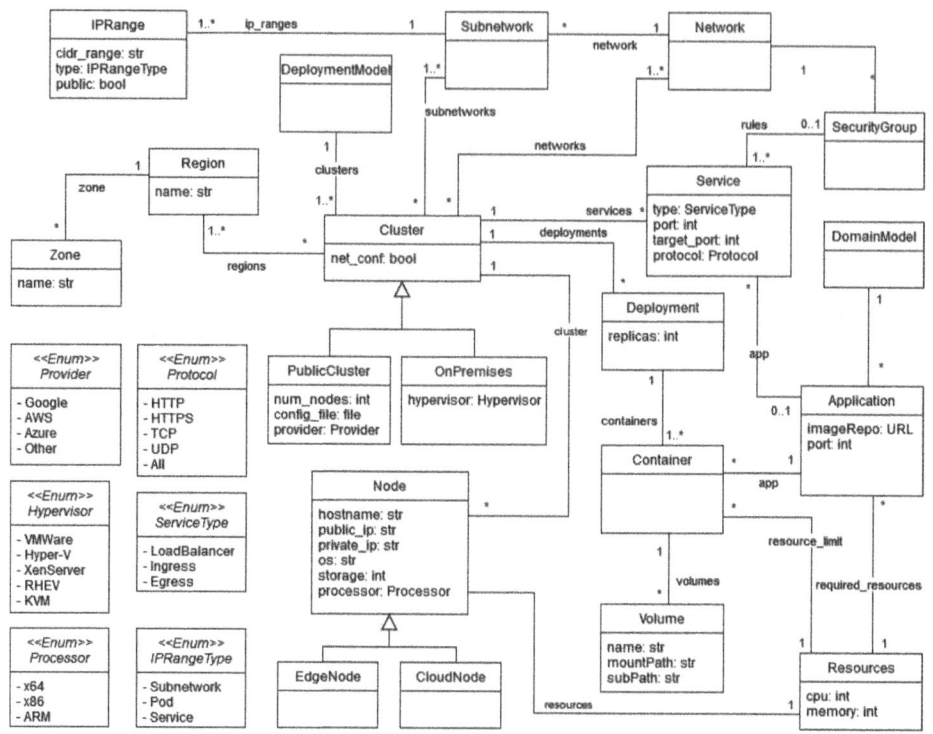

Fig. 2. Deployment Architecture Metamodel

multiple *Cluster*(s). Each *cluster* can be a *PublicCluster* or an *onPremise* one. For a *publicCluster*, the *config_file* contains the credentials and other information needed to communicate with the public cloud provider. Additionally, the architect needs to define five items for each cluster: (1) **Region**, (2) **Network**, (3) **Node**, (4) **Deployment**, and (5) **Services**. We discuss these items briefly:

As part of the region information, the architect needs to define the region/s and zones where the cluster should be created. For the network information, the architect needs to define networks, subnetworks, and IP ranges for these networks; in order to make it easier, we have a *net_conf* boolean variable to auto-generate the network and subnetworks. Regarding the node information, the OS, storage, resources, and processor for the *node*s of cluster must be defined. Each *node* can be of *CloudNode* or *EdgeNode*. As part of the fine-grained deployment information, the architect needs to specify the number of replicas of each *Deployment*; each *deployment* must have a container containing the dockerized application, volume and resource information. As for the services information, we need to provide all the services that will be attached to the clusters. Services can be load balancers, ingress or egress. Furthermore, we can also define the port mappings and security groups for the services.

This information can be entered thanks to the textual notation (i.e. concrete syntax) we have provided for the above metamodel. We have designed a grammar and a parser that converts this textual description into a B-UML model. The grammar is created using ANTLR [7], a widely used language recognition tool. We show an example of a deployment model using this grammar below but the complete grammar is available in the project repository [1].

For the purpose of exemplification throughout this study, we use the Digital Product Passport (DPP) domain, a European initiative supporting the circular economy [11]. DPP involves collecting comprehensive information on a product's composition, design, condition, and other aspects throughout its life cycle. This information can be shared with anyone interacting with the product, including manufacturers, buyers, and repair staff.

Listing 1.1 shows an excerpt of the deployment model for the DPP example, written using our defined concrete syntax. In this example, the *dpp_app* application is defined (*lines 3-5*) and packaged in the Docker image "dpp/app:latest". This image is created using the structural model, which defines the DPP domain, and is automatically uploaded to DockerHub by the backend generator. The application runs in the *dpp_container* (*lines 8-11*), with 2 replicas deployed as defined in *dpp_deployment* (*lines 14-16*). Finally, two public clusters (for Google and AWS) are defined (*lines 19-28*) to deploy the application.

Listing 1.1. Deployment model

```
1  Deployment model{
2     applications {
3         -> name: dpp_app,
4            image: "dpp/app:latest",
5            // Other info omitted for brevity purposes
6     }
7     containers {
8         -> name: dpp_container,
9            app_name: dpp_app,
10           cpu_limit: 500m,
11           memory_limit: 512Mi }
12    deployments {
13        -> name: dpp_deployment,
14           replicas: 2,
15           containers: [dppcontainer] }
16    clusters {
17        -> public_cluster
18           name: cluster_a,
19           provider: google,
20           deployments: [dpp_deployment],
21           // Other info omitted for brevity purposes
22        -> public_cluster
23           name: cluster_b,
24           provider: aws,
25           deployments: [dpp_deployment],
26    }
27    // Other concepts omitted for brevity purposes ...
28 }
```

4 Code Generation

The code generation for this BESSER extension primarily involves two generators implemented as M2T transformations using the Jinja[2] template-based engine: one for the backend and one for the Terraform code.

The Backend generator integrates multiple specialized generators from the BESSER suite (see Fig. 1) to create a comprehensive backend. It enables the creation of dynamic and scalable API endpoints using the REST API Generator, which leverages the FastAPI framework. Efficient ORM transformations for database interactions are handled by the SQLAlchemy generator, while robust data validation is ensured by the Pydantic generator, promoting data integrity and backend security. Furthermore, this generator can also package the backend as a container image and upload it to a hub for easy deployment.

On the other hand, the Terraform generator creates infrastructure as code (IaC) for provisioning and deploying the application generated using the backend generator. This code encompasses infrastructure configuration such as clusters, networks, subnets, and load balancers, as well as the deployment of software containers using Kubernetes management services like EKS[3] and GKE[4].

Using the same example from the DPP domain (introduced in Sect. 3); Fig. 3 shows the generated code for deploying the multi-cloud environment using Terraform. A set of files containing the provisioning and deployment configurations is generated for each cloud provider. The deployment process is initiated by executing the *setup.bat* file. For a detailed walkthrough and application execution, you can refer to the demo video accompanying this tool paper.

All files, code, and steps to run this example can be consulted in the BESSER example repository[5].

5 Related Work

The deployment of applications in the cloud is a well-researched topic. However, most studies do not fully integrate their solutions within a software development workflow. This is particularly true for Infrastructure as Code (IaC) tools, which, although powerful for configuring deployments, are not designed to support other phases of the application development lifecycle. Nonetheless, these tools can be incorporated within a development and deployment framework, as we propose with our low-code tool.

Some studies, such as [3,9,10], propose model-based solutions for specifying and automating cloud deployments. However, these studies do not address the modeling of multi-cloud and on-premises environments, nor do they incorporate containerization as a virtualization technology for application deployment. Other

---

[2] https://palletsprojects.com/p/jinja/.
[3] https://aws.amazon.com/eks.
[4] https://cloud.google.com/kubernetes-engine?hl=en.
[5] https://github.com/BESSER-PEARL/BESSER-examples/tree/main/examples/multi-cloud_deployment.

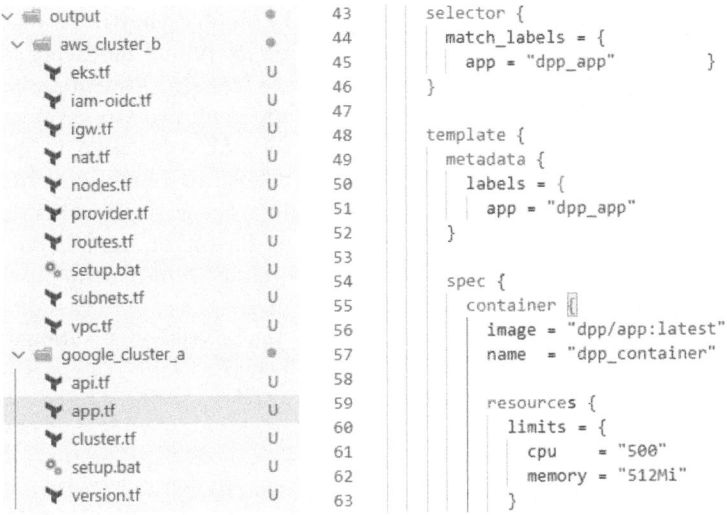

Fig. 3. Generated code for Terraform

works (such as [5, 8]) address the modeling and deploying architectures in multi-cloud environments but do not provide support for application development or containerized image support to facilitate deployment.

Commercial low-code tools also support cloud application deployment but have significant limitations in terms of configuration and customization. For instance, PowerApps and Appian typically restrict or suggest deployment to a single cloud provider, such as Azure and AWS, respectively. While some tools, like Mendix, OutSystems, and Pega, offer features for multi-cloud deployments, they limit infrastructure configuration and customization. Additionally, it is often necessary to install the low-code platform environment on a cloud provider's web server before deploying individual applications to that environment. Moreover, these tools are neither free nor open-source.

Despite the capabilities of existing solutions, there remains a gap in integrating low-code software development and multi-cloud deployment within a single tool. Our open-source low-code tool aims to bridge this gap by offering a unified framework for application development, Docker image generation, and deployment across multi-cloud environments.

6 Conclusions and Further Work

This paper extends the low-code tool, BESSER, with application packaging and multi-cloud deployment capabilities. This extension comprises several key components, including a new grammar and metamodel to specify the deployment architecture of the system, the generation of a Docker image for applications created with BESSER, and a new code generator capable of producing Terraform code for deployment across multiple cloud environments.

As part of our future roadmap, we aim to enhance the tool's usability by incorporating a new graphical notation, enabling graphical modeling of deployment architectures (e.g., using AWS Architecture Icons[6]). Additionally, we plan to enhance the capabilities of the Terraform code generator to support a wider range of cloud providers and on-premises clusters. We also plan to integrate a cloud management solution such as Google Anthos, enabling centralized management and monitoring of all clusters, including those deployed on-premises. Finally, we plan to address the continuous updating of models in runtime, enabling the integration of our tool into a continuous deployment pipeline.

Aknowledgements. This project is supported by the Luxembourg National Research Fund (FNR) PEARL program, grant agreement 16544475.

References

1. Github repo: BESSER (2023). https://github.com/BESSER-PEARL/BESSER
2. Alfonso, I., et al.: Building besser: an open-source low-code platform. In: van der Aa, H., Bork, D., Schmidt, R., Sturm, A. (eds.) BPMDS EMMSAD 2024, pp. 203–212. Springer, Cham (2024). https://doi.org/10.1007/978-3-031-61007-3_16
3. Artac, M., Borovšak, T., Di Nitto, E., Guerriero, M., Perez-Palacin, D., Tamburri, D.A.: Infrastructure-as-code for data-intensive architectures: a model-driven development approach. In: 2018 IEEE International Conference on Software Architecture (ICSA), pp. 156–15609. IEEE (2018)
4. Cabot, J.: Positioning of the low-code movement within the field of model-driven engineering. In: Proceedings of the 23rd ACM/IEEE International Conference on Model Driven Engineering Languages and Systems: Companion Proceedings, pp. 1–3 (2020)
5. Ferry, N., Chauvel, F., Song, H., Rossini, A., Lushpenko, M., Solberg, A.: Cloudmf: model-driven management of multi-cloud applications. ACM Trans. Internet Technol. (TOIT) **18**(2), 1–24 (2018)
6. Muccini, H., Sharaf, M.: Caps: architecture description of situational aware cyber physical systems. In: 2017 IEEE International Conference on Software Architecture (ICSA), pp. 211–220. IEEE (2017)
7. Parr, T.: The Definitive ANTLR 4 Reference, pp. 1–326 (2013)
8. Pham, L.M., Tchana, A., Donsez, D., Zurczak, V., Gibello, P.Y., De Palma, N.: An adaptable framework to deploy complex applications onto multi-cloud platforms. In: The 2015 IEEE RIVF International Conference on Computing & Communication Technologies-Research, Innovation, and Vision for Future (RIVF), pp. 169–174. IEEE (2015)
9. Sandobalin, J., Insfran, E., Abrahão, S.: Argon: a model-driven infrastructure provisioning tool. In: 2019 ACM/IEEE 22nd International Conference on Model Driven Engineering Languages and Systems Companion (MODELS-C), pp. 738–742. IEEE (2019)
10. Sledziewski, K., Bordbar, B., Anane, R.: A DSL-based approach to software development and deployment on cloud. In: 2010 24th IEEE International Conference on Advanced Information Networking and Applications, pp. 414–421. IEEE (2010)
11. Walden, J., Steinbrecher, A., Marinkovic, M.: Digital product passports as enabler of the circular economy. Chem. Ing. Tech. **93**(11), 1717–1727 (2021)

---

[6] https://aws.amazon.com/architecture/icons/.

Doctoral Symposium

Reference Architecture of MLOps Workflows

Faezeh Amou Najafabadi

Vrije Universiteit Amsterdam, Amsterdam, The Netherlands
f.amou.najafabadi@vu.nl

Abstract. The rapid growth in the adoption of Machine Learning Operations Workflows (MLOps WFs) has given rise to the development of numerous guidelines and tools aimed at supporting the creation and management of these WFs. However, MLOps stakeholders continue to encounter challenges in employing these guidelines and tools. Firstly, there is a lack of consensus on the standard implementation of MLOps. Secondly, the current tools only support one or a number of components within MLOps workflows, complicating their integration into end-to-end WFs. Furthermore, the tendency towards automation in MLOps has led to uncertainty about the optimal level of human involvement, raising concerns about whether complete automation is the ideal approach. Responding to these issues, our goal in this research is to aid the MLOps WF stakeholders by providing a comprehensive reference architecture, that can be consulted as a basis of consolidated knowledge and experience in designing and managing MLOps WFs.

Keywords: Reference Architecture · Machine Learning Operations · MLOps · MLOps Workflow · MLOps Process

1 Introduction

With the increase in the volume of data and the complexity of Machine Learning (ML) pipelines, Machine Learning Operations (MLOps) is introduced as a solution to automate and optimize the end-to-end ML pipeline in production [8]. In the traditional ML-based systems, the ML models are developed, tested, and integrated into the embedding software to serve predictions. However, in most cases, it is reported that even though the tested models perform well in the staging environment, they are not successfully deployed in the production environment [1]. To tackle this, MLOps was introduced to address the successful deployment of ML models in production [6]. As opposed to traditional ML-based systems (that only host a trained and tested ML model), MLOps Workflows (WFs) consist of multiple stages that correspond to building, deploying, and monitoring the traditional ML pipelines. In this research, our focus is on the MLOps WFs, i.e. the paradigm that facilitates operationalizing and serving ML pipelines in production.

MLOps, as a successor of DevOps [5], introduces several benefits to traditional ML-based systems. By the integration of CI/CD, MLOps provides continuous unit testing of ML models integrated into the embedding software and continuous delivery of the models to the production environment [6]. This enhances time to production. Moreover, by the introduction of model monitoring, MLOps prevents the degradation of ML predictions in inference. This is implemented through the continuous training pipeline (CT).

Despite the introduced benefits, the creation of MLOps WFs is not a straightforward task. In this research, we address three major problems that are faced when conducting MLOps WFs. First, with the lack of a standard definition of MLOps WFs [12], architecting an end-to-end MLOps Workflow (WF) requires multiple iterations of trial and error. Moreover, there exist several tools supporting the MLOps WFs, e.g. MLFlow[1], MetaFlow[2], ZenML[3], and Prefect[4]. However, the existing tools either support only certain stages of the MLOps WF, or if they support the end-to-end WF, practitioners tend to use them only for certain stages. As a result, selecting the best combination among the available options that (a) suits the needs of the MLOps WF stakeholders (i.e. data scientists, MLOps engineers, software engineers, and operations teams), and (b) provides quality attributes e.g. performance and reliability, has been problematic [4]. Lastly, in MLOps WF architectures, optimization is usually intertwined with automation. With the introduction of Automated Machine Learning (AutoML) and similar concepts in the MLOps WFs, the main aim has been to eliminate human intervention within the WFs which is not always the optimal solution [2]. To address this, we plan to integrate the concept of *human-in-the-loop (HITL)* [14] within the MLOps WFs. Through the HITL integration, we can for instance provide the possibility for stakeholders to interrupt the retraining pipeline and prioritize time-to-production over model accuracy during the retraining of an ML model. This provides the stakeholders the possibility of optimizing the MLOps WF based on their emerging needs during the production phase.

Our main goal in this PhD research is to provide MLOps WF stakeholders with a Reference Architecture (RA) [15] that helps them design the architecture of MLOps WFs, integrate the available tools effectively, and determine the trade-off of automation versus human intervention and the points of the human intervention within the WF. Our envisioned RA comprises three main elements: (i) architecture blueprints; (ii) a comprehensive set of guidelines regarding best and worst practices, patterns, antipatterns, tool integration, and human intervention within the automated WFs; and (iii) common stakeholder requirements.

---

[1] https://mlflow.org.
[2] https://metaflow.org.
[3] https://www.zenml.io.
[4] https://www.prefect.io.

2 Related Work and Problem Statement

MLOps definition and architecture. Several scientific papers try to disambiguate the definition of MLOps. Kreuzberger et al. [9] define MLOps under four main aspects including architecture. In their preprint, Kumara et al. [10] provide a layered architecture for MLOps focusing on the requirements that an MLOps environment needs to fulfill by reviewing the gray literature. Although this work covers aspects related to the structure and stakeholders of an RA for MLOps, it does not provide a comprehensive set of guidelines regarding practices, patterns, tool integration, and stakeholder involvement that we intend to provide. Raffing et al. [17] provide an RA in the manufacturing domain. John et al. [6] describe the different stages that companies go through in adopting MLOps practices.

Despite the various studies regarding defining MLOps and providing architectures, "MLOps remains at its initial stages" [11] with a lack of standard domain-independent references that practitioners can follow to create MLOps WFs that suit their needs. Moreover, the aforementioned architectures are proposed either for specific domains [17] or only cover specific aspects that a reference architecture entails [9, 10].

MLOps tools. Symeonidis et al. [19] focus on presenting different tools and their usefulness in the MLOps environment, categorizing the tools based on their functionalities. The authors also provide general guidelines for tool selection and integration. Kolltveit and Li [8] focus on the operationalization of ML models concerning tools and infrastructure. As a contribution, their work describes how different tools and infrastructures are used in operationalizing ML models. Recupito et al. [18] provide an overview of the most common tools supporting an MLOps pipeline and their features. In a follow-up study [13], the authors provide a graphical mapping of the identified MLOps tools and DevOps phases. Moreover, they provide valuable insights into combining the tools in an end-to-end ML pipeline and potential incompatibilities between the tools.

While the existing literature regarding the tools that support MLOps WFs, provides a comprehensive list of the tools and general guidelines for combining them in an end-to-end MLOps WF, they do not cover detailed guidelines describing the efficient combination of specific tools to create an end-to-end MLOps WF.

Human in the loop. Wu et al. [21] conducted a survey on existing works on human-in-the-loop (HITL) from a data perspective, i.e. using human expertise in the data processing pipeline of the ML WFs. Gebauer et al. [3] present a prototype system that integrates HITL in privacy policy annotation generated by ML models. Ostheimer et al. [7] provide design principles for developing machine learning algorithms that include the concept of HITL.

While existing works describe the impacts of engaging humans in different phases of ML pipelines and propose guidelines on how to design hybrid-learning pipelines, to the best of our knowledge, no study addresses optimizing ML deployment through the definition of the degree of human intervention versus complete automation and the identification of points of human intervention in an automated MLOps WF.

To complement the above works and address the identified gaps, we aim to provide a comprehensive RA of MLOps WFs in our research, which can be consulted as a basis for consolidated knowledge about MLOps WF architectures. Proposing this RA entails addressing the following problems P1-P3.

P1:There is still no common understanding among researchers and experts on how MLOps should be designed and implemented. Despite the available guidelines and references, MLOps WF stakeholders in different domains still need to spend considerable time interpreting how to design an MLOps WF architecture that suits their requirements. In their systematic review of 60 primary studies, Mboweni et al. [12] "did not find evidence of any common understanding amongst scholars and experts on how MLOps should be implemented and institutionalized across the industry to create a common vision."

P2: A considerable number of tools supporting the different stages of MLOps WFs already exist. However, there are still no specific guidelines on how to best integrate these tools for creating an end-to-end MLOps WF. The availability of different tools for each stage or partial support of the MLOps WF leads to numerous combinations of the available options to create an end-to-end WF. Due to a missing reference on how to best integrate the available tools [10] in accordance with their needs, MLOps WF stakeholders in different sectors must individually go through all the stages and invest considerable time and effort in creating their WFs by choosing among these tools.

P3: The extent of human involvement in the loop, in contrast to the degree of automation, remains unclear in the context of MLOps WFs. With the automation of the processes and use of AutoML, practitioners focus on eliminating user interaction as much as possible in ML pipelines. While this approach can be beneficial, preserving human intervention [20] in MLOps WFs leads to notable benefits in many cases e.g. experts intervening in automated ML training pipeline to prioritize time-to-production over model accuracy.

3 Research Questions and Proposed Approach

In order to address the problems P1-P3 in Sect. 2, we specify our research plan as depicted in Fig. 1 formulating our research questions as follows:

RQ1: *What is a comprehensive reference architecture of MLOps workflows?*
To address this research question, we examine the following sub-questions.

- **RQ1.1:** *How are the MLOps WF architectures defined in scientific literature?*
  To provide a domain-agnostic RA we need to establish a thorough understanding of how MLOps WF architectures are defined in the state of the art. To this end, we conduct a systematic mapping study [16] to elicit MLOps architectures from structural and process views. To provide the structural

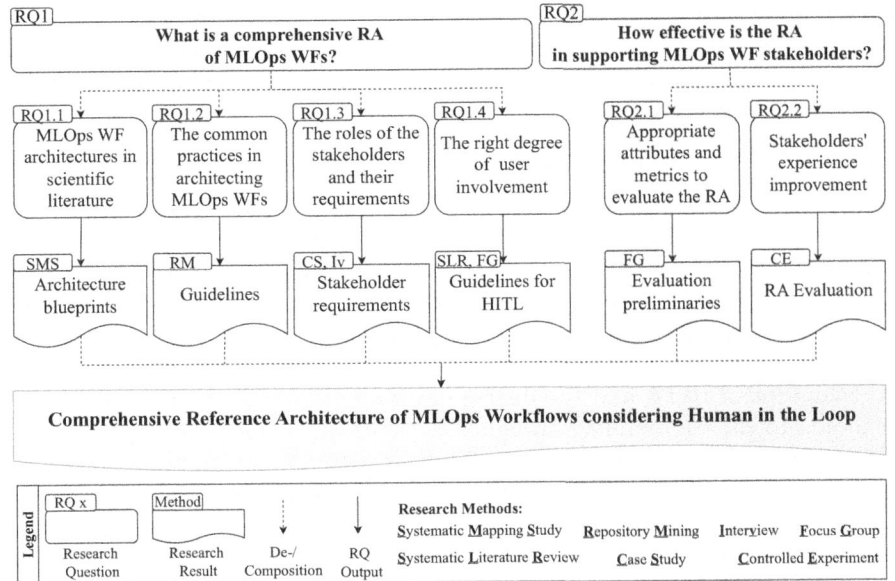

Fig. 1. The Proposed Research Plan

view of an MLOps WF architecture, we establish the architecture components and their relationships that can be referred to when implementing an end-to-end MLOps WF. Similarly, to provide the process view, we establish activities that are performed in an end-to-end MLOps WF. The synthesized structural and process views eventually form the blueprints in our envisioned RA.

- **RQ1.2:** *What are the common practices when architecting MLOps WFs?*
  To provide guidelines and present practices in the RA we establish the current best and worst practices leading to patterns and antipatterns while designing the architecture of an MLOps WF. To this end, we conduct repository mining to elicit information encompassing practitioners' experiences and challenges by extracting questions, comments, and reviews on social platforms and software repositories. The expected result is a set of documented guidelines, also including guidelines for tool integration that serve MLOps WF stakeholders to build their WFs based on the expertise of leading practitioners, thereby minimizing reliance on trial and error methods when architecting MLOps WFs.

- **RQ1.3:** *What are the roles of the stakeholders in an MLOps WF, and what are the requirements per stakeholder?*
  One of the crucial elements of an effective RA is stakeholder requirements in different domains and scenarios. To provide this, we elicit the potential MLOps WF stakeholders and their requirements throughout the entire WF. The results provide a mapping between stakeholders, their requirements, and

the stages of the MLOps WF. We approach this research sub-question by conducting a descriptive case study and practitioner interviews. In particular, we plan to observe the stakeholders throughout the entire MLOps WF in different domains to elicit their roles. Consequently, we interview potential stakeholders of each stage of the MLOps WF to identify their requirements.

- **RQ1.4:** *How to define the right degree of automation versus user involvement in an end-to-end MLOps WF?*

By this sub-question, we aim to identify the best practices for integrating human intervention into the automated pipelines to obtain an efficient MLOps WF. We approach this research sub-question by conducting a mixed methods research comprised of a systematic literature review and discussions within a focus group to extract the established state of the art of this trade-off between the concept of HITL and the degree of automation and the insights that we can add to the community through the results of the focus group discussions. Drawing upon the results of RQ1.3, we also create a mapping between the responsible users and each point of human intervention within the automated pipelines of MLOps WFs. The expected results are documents of guidelines to consult practitioners to take design decisions about HITL points of action.

Synthesizing the results of subquestions RQ1.1 - RQ1.4 provides us with our envisioned RA, which we evaluate through RQ2.

RQ2: *How effective is the proposed RA in supporting MLOps WF stakeholders?*
To evaluate our RA, we address the following sub-questions.

- **RQ2.1:** *What are the appropriate attributes and related metrics for evaluating the effectiveness of the proposed RA from different stakeholders' perspectives?*

To enhance stakeholders' experience by using the proposed RA, we need to elicit the attributes and metrics that affect each stakeholder's experience in architecting and implementing an MLOps WF. We plan to form focus groups through which we establish the appropriate attributes and metrics for the evaluation of the effectiveness of our proposed RA for different stakeholders. The results help us identify suitable attributes and metrics that complement the existing ones for evaluating software and systems architectures, and current reference architectures for different domains and other technologies (e.g. IoT, AI, etc.). The expected outcome builds the foundations for RQ2.2.

- **RQ2.2:** *To what extent is the stakeholders' experience improved by using the proposed RA?*

To evaluate the proposed RA, we measure the stakeholders' experience improvement by comparing their experience designing concrete MLOps WF architectures with or without using our RA. We evaluate the experience by employing the identified attributes and metrics from RQ2.1 and conducting a controlled experiment. For the controlled experiment, we plan to involve junior MLOps stakeholders who do not yet have architecture experience and would need training. Among the participants, we engage a test group (provided with the RA) and a control group (who do not have an RA) and define tasks for them regarding architecting an MLOps WF. The experience of the individuals within the two groups is then evaluated through the attributes and metrics of RQ2.1. The expected results include a description of user experience and RA evaluation as well as points of enhancement for the RA. However, we foresee the challenge of controlling the independent variables in experiments that involve human experts which can be a complex task and needs a thorough experiment design.

In general, we plan to conduct two iterations of our research. The first iteration results in the proposal (RQ1) and evaluation (RQ2) of our RA, and the second iteration results in the enhancement of our initial RA and evaluation of the improved version. The research methods in the second iteration will be different from the methods described here, to assist the enhancement of the RA, e.g. thorough scoping and literature reviews will not be necessary anymore. However, the exact research design for the second iteration remains an open question at this point depending on the results of the first iteration of the research.

4 Conclusion

In this paper, we present our research plan in terms of research questions, the related methodology, and expected results. To achieve the goal of this research, namely providing MLOps WF stakeholders with a comprehensive RA, we propose undertaking a mix of research methodologies including scoping and literature reviews, repository mining, case studies, interviews, focus groups, and controlled experiments involving practitioners to collect and synthesize data that leads us to the artifact of our research, which is a proposed and evaluated RA of MLOps WFs. Our envisioned RA of MLOps WFs, once established, can be consulted by researchers and practitioners to design MLOps WF architectures based on their requirements.

Acknowledgments. This research is supported by ExtremeXP, a project co-funded by the European Union Horizon Programme under Grant Agreement No. 101093164.

References

1. Algorithmia: state of enterprise machine learning. Tech. rep., Algorithmia (2020)
2. Brauner, P., Ziefle, M.: Why consider the human-in-the-loop in automated cyber-physical production systems? two cases from cross-company cooperation. In: 2019 IEEE 17th International Conference on Industrial Informatics (INDIN). vol. 1, pp. 861–866 (2019). https://doi.org/10.1109/INDIN41052.2019.8972142
3. Gebauer, M., Maschhur, F., Leschke, N., Grünewald, E., Pallas, F.: A 'human-in-the-loop'approach for information extraction from privacy policies under data scarcity. In: 2023 IEEE European Symposium on Security and Privacy Workshops (EuroS&PW), pp. 76–83. IEEE Computer Society, Los Alamitos, CA, USA (jul 2023). https://doi.org/10.1109/EuroSPW59978.2023.00014
4. Idowu, S., Strüber, D., Berger, T.: Asset management in machine learning: A survey. In: IEEE/ACM 43rd International Conference on Software Engineering: Software Engineering in Practice (ICSE-SEIP), pp. 51–60 (2021). https://doi.org/10.1109/ICSE-SEIP52600.2021.00014

5. Jabbari, R., bin Ali, N., Petersen, K., Tanveer, B.: What is devops? a systematic mapping study on definitions and practices. In: Proceedings of the scientific workshop proceedings of XP2016, pp. 1–11 (2016). https://doi.org/10.1145/2962695.2962707
6. John, M.M., Olsson, H.H., Bosch, J.: Towards mlops: A framework and maturity model. In: 2021 47th Euromicro Conference on Software Engineering and Advanced Applications (SEAA), pp. 1–8 (2021). https://doi.org/10.1109/SEAA53835.2021.00050
7. Ostheimer, J. Chowdhury, S., S.I.: An alliance of humans and machines for machine learning: Hybrid intelligent systems and their design principles. Technol. Soc. (2021). https://doi.org/10.1016/j.techsoc.2021.101647
8. Kolltveit, A.B., Li, J.: Operationalizing machine learning models - a systematic literature review. In: IEEE/ACM 1st International Workshop on Software Engineering for Responsible Artificial Intelligence (SE4RAI), pp. 1–8 (2022). https://doi.org/10.1145/3526073.3527584
9. Kreuzberger, D., Kühl, N., Hirschl, S.: Machine learning operations (MLOPS): Overview, definition, and architecture. IEEE Access **11**, 31866–31879 (2023). https://doi.org/10.1109/ACCESS.2023.3262138
10. Kumara, I., Arts, R., Di Nucci, D., Van Den Heuvel, W.J., Tamburri, D.A.: Requirements and reference architecture for mlops: Insights from industry. Authorea Preprints (2023). https://doi.org/10.36227/techrxiv.21397413.v1
11. Lima, A., Monteiro, L., Furtado, A.P.: MLOPS: practices, maturity models, roles, tools, and challenges-a systematic literature review. ICEIS **1**, 308–320 (2022). https://doi.org/10.5220/0010997300003179
12. Mboweni, T., Masombuka, T., Dongmo, C.: A systematic review of machine learning devops. In: 2022 International Conference on Electrical, Computer and Energy Technologies (ICECET), pp. 1–6. IEEE (2022). https://doi.org/10.1109/ICECET55527.2022.9872968
13. Moreschini, S.: Toward end-to-end MLOPS tools map: a preliminary study based on a multivocal literature review. arXiv preprint arXiv:2304.03254 (2023). https://doi.org/10.48550/arXiv.2304.03254
14. Mosqueira-Rey, E., Hernández-Pereira, E., Alonso-Ríos, D., Bobes-Bascarán, J., Fernández-Leal, Á.: Human-in-the-loop machine learning: a state of the art. Artif. Intell. Rev. **56**(4), 3005–3054 (2023). https://doi.org/10.1007/s10462-022-10246-w
15. Nakagawa, E.Y., Antonino, P.O.: Reference Architectures for Critical Domains. Springer Cham (2023). https://doi.org/10.1007/978-3-031-16957-1
16. Petersen, K., Vakkalanka, S., Kuzniarz, L.: Guidelines for conducting systematic mapping studies in software engineering: An update. Inf. Softw. Technol, p. 1-18 (2015). https://doi.org/10.1016/j.infsof.2015.03.007
17. Raffin, T., Reichenstein, T., Werner, J., Kühl, A., Franke, J.: A reference architecture for the operationalization of machine learning models in manufacturing. Procedia CIRP **115**, 130–135 (2022). https://doi.org/10.1016/j.procir.2022.10.062
18. Recupito, G., et al.: A multivocal literature review of MLOPS tools and features. In: 2022 48th Euromicro Conference on Software Engineering and Advanced Applications (SEAA), pp. 84–91. IEEE (2022). https://doi.org/10.1109/SEAA56994.2022.00021
19. Symeonidis, G., Nerantzis, E., Kazakis, A., Papakostas, G.A.: MLOPS - definitions, tools and challenges. 2022 IEEE 12th Annual Computing and Communication Workshop and Conference (CCWC), pp. 0453–0460 (2022). https://doi.org/10.1109/CCWC54503.2022.9720902

20. Wang, J., Guo, B., Chen, L.: Human-in-the-loop machine learning: a macro-micro perspective. arXiv preprint arXiv:2202.10564 (2022). https://doi.org/10.48550/arXiv.2202.10564
21. Wu, X., Xiao, L., Sun, Y., Zhang, J., Ma, T., He, L.: A survey of human-in-the-loop for machine learning. Futur. Gener. Comput. Syst. **135**, 364–381 (2022). https://doi.org/10.1016/j.future.2022.05.014

Evaluating the Effect of Team Ownership of Microservices: Strategies for Balancing Decoupling, Coordination, and System Cohesion

Noman Ahmad(✉)

Oulu University, Oulu, Finland
noman.ahmad@oulu.fi

Abstract. The increasing use of microservices architecture demonstrates how its emphasis on technological limits and modularity may improve software design and delivery. Nonetheless, difficulties still exist about ownership of microservice, service integration, and team decoupling. The PhD research aims to investigate team ownership in microservices and suggest ways to achieve a balance between decoupling, coordination, and system cohesiveness. The study will specifically address three important questions: Initially, it will look into methods of achieving a balance between the needs for coordination and service integration to maintain system cohesion and team decoupling. Secondly, I aim to investigate the impact of clear service ownership on the efficiency and quality of service development. Moreover, the research aim is to examine how different skill combinations, role distribution, team sizes and personalities affect the performance of microservice system through an analysis of team composition.

Keywords: Microservices · Team Ownership · Decoupling

1 Introduction and Related Work

Microservice is gaining popularity in software service design and delivery. They make up a method of designing software and systems architecture that emphasizes technological boundaries while expanding upon the well-known modularization idea [9]. The concept of decoupled teams is supported by microservices architecture, which lets each team focus on specific services [14]. Each team is in charge of its own service from conception to implementation, and the "one team working on one service" concept can help minimize dependencies between teams [2,9,14]. Even though teams are decoupled, their services sometimes need to interact to form a complete system, which creates an anti-pattern, and it is also an issue to find the ownership of microservice between the teams. All of the microservices in a system are developed, deployed, and maintained by a single team. It violates the principles of microservices architecture, which supports autonomy and decentralizes teams. Bad communication or inconsistency in alignment can cause integration issues, inconsistent APIs, or data mismatches [1,4,6,9,17].

The previous research investigated organizational coupling, such as [13] performed research to evaluate how any two microservices are coupled by the cross-service contribution behaviors of the developers. Li et al. [13] performed a case study on one project, which is adapted in this Study to get the results. So, the research on an organizational level with more microservices projects involved can also help my research work in determining the approach to balance the decoupling and system cohesion [13]. In terms of team ownership, there are some studies, such as d'Aragona et al. [2] that look at the practice of having one developer per microservice in open-source software (OSS) projects, specifically looking at microservice ownership. According to an empirical study conducted on 38 microservice-based open-source software (OSS) projects, their results show that allocating one microservice per developer is rarely implemented [2]. Li et al. [12] performed an analysis of microservices projects' organizational structures based on contributor collaboration. Their Study shows how network analysis can be used to efficiently visualize and evaluate collaboration arrangements, highlighting the centrality and modularity of contributions within the project. According to Conway's Law, any system designed by an organization will have a structure similar to its communication structure, it was discovered that microservices projects frequently feature a decentralized organization [12]. There is also a research study [11] about developer collaboration in microservices to know the patterns, and it highlights that there is potential for improvement in collaboration structures by identifying multiple important developer communities that are not necessarily in line with the ideal microservice architecture inside a sample microservice project. The study [11] also shows that the personality traits of the developers have an essential effect on team dynamics and project performance. This Study was also in the early development stage, so there is a potential for further development and validation [11].

So, there is research work performed related to the organizational level but there is not much work done previously specifically on investigating clear team ownership in microservice, which provides a balanced approach to tackle decoupling, coordination, and cohesion. The PhD aims to investigate team ownership in microservices, show various team structures, specifically in terms of their coupling, and provide recommendations for balancing decoupling, coordination, and system cohesion. Moreover, the plan is to visualize the results using Grafana or any other visualization method.

2 Objectives

In a microservices architecture, clear ownership is crucial as each service can be seen as a separate product with its lifecycle, requiring specific domain knowledge. To address the aforementioned issue, I formulated three research questions (RQs) as the basis of my PhD research:

RQ1: How to balance team decoupling and the requirements for integration, coordination, and overall system cohesion?

RQ2: How a clearly defined ownership impacts the efficiency and quality of service development?

RQ3: How do different team compositions affect the performance of a microservice system, particularly in detecting the organization coupling issues between teams and finding solutions to mitigate such issues to enhance the performance?

3 Approach and Expected Results

- The results for RQ1 could possibly be a collection of recommendations and the best approach to balance the coupling between the teams. For RQ1, a similar approach can be used by Li et al. [13] because they proposed using organization coupling between microservices and evaluated coupling between any two microservices caused by the developers' cross-service contribution behavior. The case study performed by Li et al. [13] was based on only one project in their research to get the results. Their study [13] involves identifying the developers who contributed to the microservice and identifying the contributor team for each microservice. They also calculated the contribution of each developer on each microservice. So, the research involving more projects related to microservice on an organizational level can help my research work in determining the approach to balance the decoupling and system cohesion [13]. The idea is also to visualize the results similar to this study from Cerny et al. [5] by adding another organizational layer.
    - The research can help researchers for future studies in a microservices architecture. Researchers can use these results as a benchmark for measuring new models and methods.

- For RQ2, finding the owner in microservice projects and utilizing software solutions that track and map ownership of different microservices architecture components. This can help determine the team or individual responsible for each service as well as the interdependencies between them. A comparison between teams with clear ownership and those with less clear ownership can also help to evaluate the impact on efficiency and quality of microservice by clear team ownership. A study by Bird et al. [3] about code ownership on software quality shows the proportion of work on each component, such as the total number of commits during the development life cycle. So, similar approach can be utilized to check the developers' proportion of work on each microservice and determine ownership.
    - It can contribute to the field by delivering evidence-based practices that others can confidently adopt, such as researchers for future research and organizations for implementing microservices between different teams.

– For RQ3, statistical analysis can be performed to analyze quantitative data such as team composition, including team size, skill diversity, personalities, and role distribution. This analysis will help understand the impact of different team compositions on the performance of microservice systems, particularly in detecting and mitigating organizational coupling issues. There is a study about developer personalities for optimizing collaboration in microservice projects by Li et al. [11]. Our results will provide knowledge about team composition's impact on the performance of the microservice system.
  - The outcomes will not just help researchers but also provide insights to organizations on how they can structure teams around microservices, which can help optimize performance, which is critical in fast-paced industries.

4 Research Methods

For RQ1, Case study [16] analysis similar to Li et al. [13] on the organizational level to find an approach that best balances integration requirements with decoupling. A comparative study can be performed for RQ2 to evaluate the development efficiency and service quality of projects with well-defined ownership structures compared to those with less clear ownership. It will help to evaluate the clear impact of ownership on quality and service development. In RQ3, Statistical analysis can be performed to analyze quantitative data [8] such as data on team composition, such as team size, skill diversity, and role distribution. The dashboard for visualization can be used. Moreover, other methods can also be utilized in the future depending on requirements.

5 Research Ethical Issues

This research will involve human participants. We will ensure that they get the full information about the use and purpose of the research and that their consent is requested. Confidentiality and Privacy will also be taken into account during the research. GDPR will be considered when collecting that sort of data.

6 Implementation

The research work will be performed in three stages.

– Stage 1
  - A Literature review will be performed about microservices to get the existing knowledge. The guidelines provided by Petersen et al. [15] and Kitchenham et al. [10] will be utilized.
  - Identification of suitable microservices dataset available such as utilizing d'Aragona et al. [7] dataset.

- Collection of the data from public repositories, like GitHub, focusing on metrics relevant to microservices, including performance metrics, logs, and error rates.

– Stage 2
- Performing initial analysis, such as doing statistical analysis to identify the correlations between various team structures and metrics.
- Selection of case study by identifying case studies of different organizations that utilize microservices to comprehend how various team ownership and service composition models impact performance. Analyzing how organizations balance the decoupling of teams with the need for integration and effective communication.
- Development of models that can predict how team size, composition, and skill diversity will affect the microservices performance.
- Working on the dashboard for visualizing the results.

– Stage 3
- Performing final analysis.
- Testing and getting feedback about the dashboard.
- Performing final adjustments.
- Documentation of findings, model methodologies, and recommendations for future research and industry adoption.

7 Research Team, Supervision, and Collaboration

This PhD work will be performed in the M3S research Unit and under the supervision of Davide Taibi from the University of Oulu. The work will also be performed with other M3S group colleagues and our team will also collaborate with the team in the University of Helsinki. This research work will be performed in the context of the project MuFano(Multimodal Fusion-based Anomaly Detection for Improving Microservice-based System).
The external supervision of my research will be done by Andrea Janes from the Free University of Bozen/Bolzano. The research work will be published in Software architecture or microservices-related conferences such as Saner, ICSE, ECSA, ICSA, etc.

8 Critical Reflection

The approach of quantitative analysis, comparative studies, and case studies provides a strong framework for addressing research objectives. There can be some challenges, as one of the biggest challenges can be getting valid data such as projects in which it's difficult to get logs. To overcome this challenge, the plan is to use open-source microservice projects such as those mentioned in d'Aragona et al. [7] and filter those projects according to the requirement. For industrial projects, we need to find collaboration with the industry and utilize their data.

Moreover, the dashboard can be designed and implemented in Grafana. Data quality inconsistencies can be an issue for the results, especially for the open-source projects where documentation and data logging may not be standardized. Moreover, These difficulties also show how important it is to carefully choose which projects to perform the research. Selecting research projects that have high standards for data management and transparency will be essential to reducing these problems and ensuring the validity and reliability of the study findings. By utilizing the full potential of the selected approaches, this careful selection procedure will enable more relevant and accurate interpretations that have the potential to make a major contribution to the field.

9 Conclusion

In this PhD study, team ownership in microservices architecture will be investigated, with a focus on decoupling, coordination, and system cohesion. The research will show the effect of team ownership of microservice using a combination of research methods and techniques. Results will also be visually represented on a proposed dashboard. To guarantee accurate results, the research will carefully choose projects addressing issues beforehand, like data accessibility and quality. The goal of this work is to make a substantial contribution to the field of software architecture research by offering practical insights and standards for future studies.

References

1. Aksakalli, I.K., Çelik, T., Can, A.B., Tekinerdoğan, B.: Deployment and communication patterns in microservice architectures: a systematic literature review. J. Syst. Softw. **180**, 111014 (2021)
2. Amoroso d'Aragona, D., Li, X., Cerny, T., Janes, A., Lenarduzzi, V., Taibi, D.: One microservice per developer: is this the trend in OSS? In: European Conference on Service-Oriented and Cloud Computing, pp. 19–34. Springer (2023). https://doi.org/10.1007/978-3-031-46235-1_2
3. Bird, C., Nagappan, N., Murphy, B., Gall, H., Devanbu, P.: Don't touch my code! examining the effects of ownership on software quality. In: Proceedings of the 19th ACM SIGSOFT Symposium and the 13th European Conference on Foundations of Software Engineering, pp. 4–14. ESEC/FSE '11, Association for Computing Machinery, New York, NY, USA (2011). https://doi.org/10.1145/2025113.2025119
4. Cerny, T., Abdelfattah, A.S., Al Maruf, A., Janes, A., Taibi, D.: Catalog and detection techniques of microservice anti-patterns and bad smells: a tertiary study. J. Syst. Softw. **206**, 111829 (2023)
5. Cerny, T., Abdelfattah, A.S., Yero, J., Taibi, D.: From static code analysis to visual models of microservice architecture. Cluster Comput. 1–26 (2024)
6. Daniel, J., Wang, X., Guerra, E.: How to design future-ready microservices? Analyzing microservice patterns for adaptability. In: Proceedings of the 28th European Conference on Pattern Languages of Programs, pp. 1–7 (2023)

7. d'Aragona, D.A., et al.: A dataset of microservices-based open-source projects. In: Proceedings of the 21st International Conference on Mining Software Repositories (2024)
8. Ghafar, Z.N.: Evaluation research: a comparative analysis of qualitative and quantitative research methods. Middle East Res. J. Linguist. Lit. **3**(02), 25–32 (2023). https://doi.org/10.36348/merjll.2023.v03i02.003
9. Jamshidi, P., Pahl, C., Mendonça, N.C., Lewis, J., Tilkov, S.: Microservices: the journey so far and challenges ahead. IEEE Softw. **35**(3), 24–35 (2018)
10. Kitchenham, B., Brereton, P., Budgen, D., Turner, M., Bailey, J., Linkman, S.: Systematic literature reviews in software engineering-a systematic literature review. Inf. Softw. Technol. **51**, 7–15 (2009). https://doi.org/10.1016/j.infsof.2008.09.009
11. Li, X., Calefato, F., Lenarduzzi, V., Taibi, D.: Toward collaboration optimization in microservice projects based on developer personalities. In: International Conference on Software Architecture (ICSA) (2024)
12. Li, X., Abdelfattah, A.S., Yero, J., d'Aragona, D.A., Cerny, T., Taibi, D.: Analyzing organizational structure of microservice projects based on contributor collaboration. In: 2023 IEEE International Conference on Service-Oriented System Engineering (SOSE), pp. 1–8. IEEE (2023)
13. Li, X., d'Aragona, D.A., Taibi, D.: Evaluating microservice organizational coupling based on cross-service contribution. In: International Conference on Product-Focused Software Process Improvement, pp. 435–450. Springer (2023). https://doi.org/10.1007/978-3-031-49266-2_30
14. Newman, S.: Building microservices: designing fine-grained system. O'Reilly Media, Inc., California, pp. 2 (2015)
15. Petersen, K., Feldt, R., Mujtaba, S., Mattsson, M.: Systematic mapping studies in software engineering. In: Proceedings of the 12th International Conference on Evaluation and Assessment in Software Engineering **17** (2008)
16. Runeson, P., Höst, M.: Guidelines for conducting and reporting case study research in software engineering. Empir. Softw. Eng. **14**, 131–164 (2009)
17. Waseem, M., Liang, P., Shahin, M., Di Salle, A., Márquez, G.: Design, monitoring, and testing of microservices systems: the practitioners perspective. J. Syst. Softw. **182**, 111061 (2021)

Technical Debt and Software Quality in Cloud-Native Applications

Ruoyu Su[✉]

M3S, University of Oulu, Oulu 90570, Finland
ruoyu.su@oulu.fi

Abstract. With the rapid development of cloud computing technologies, cloud-native applications play an increasingly important role in software architecture. However, as applications iterate and evolve, technical debt (TD) can be a critical issue in reducing software quality. Such issues require careful attention and maintenance to solve. My doctoral research aims to investigate the various perspectives of the TDs hidden in cloud-native applications that could possibly influence the quality of the software architecture. The perspectives include the patterns, identification, evaluation, and management strategies of TD in cloud-native applications. This research can help development teams better cope with the challenges posed by TD and improve the quality and maintainability of cloud-native applications on the software architecture.

Keywords: Technical debt · Cloud-native · Software architecture · Software quality · Microservices

1 Introduction

Cloud-native applications have become the mainstream in software architecture with the rapid development and popularity of cloud computing technology [2]. A cloud-native application is a program designed for a cloud computing architecture [7]. These applications are run and hosted in the cloud and are designed to capitalize on the inherent characteristics of a cloud computing software delivery model [5]. Cloud-native applications mainly use a microservice architecture, and this architecture efficiently allocates resources to each service that the application uses, making the application flexible and adaptable to a cloud architecture [9]. However, software quality problems are increasingly highlighted as cloud-native applications iterate and evolve [8]. Among them, technical debt (TD) refers to the compromises made during the software development process for quick delivery, resulting in subsequent problems that require more time and resources to solve [12]. In cloud-native applications, TD may involve many aspects, such as architectural design, code quality, test coverage, security, etc. [18]. These TDs not only affect the quality and stability of the application but also increase the cost and risk of subsequent maintenance [11].

Therefore, it is crucial to conduct in-depth research and exploration of TD in cloud-native applications. This doctoral research aims to provide insights into the patterns and types, identification, evaluation, and management strategies of TD in cloud-native applications to help development teams better address the challenges posed by TD and improve the quality and maintainability of their applications. Through this research, we expect to provide new theoretical and practical guidance on the issue of TD in the field of software engineering and promote continuous improvement and innovation in cloud-native applications.

The remainder of this paper is organized as follows: Sect. 2 presents an overview of related work. In Sect. 3, we define the objectives of our work and Sect. 4 outlines the research methodology adopted in the study. Section 5 concludes the working plan and expected results, and Sect. 6 is a critical reflection on the work. Finally, Sect. 7 provides the conclusion.

2 Related Work

Many studies have been published on TD in microservice architectures. Firstly, there is some secondary research on TD in microservices. Villa et al. [17] conducted a systematic mapping study to investigate the types and characteristics of technical debt in systems with microservices architecture, deepening the study of these types of technical debt systems based on microservices. Tapia and Gaona [13] conducted a systematic literature review to identify key criteria and challenges in assessing the quality of microservices, where technical debt is a critical aspect.

There are also many case studies based on industry investigating the quality and maintenance of TD in microservices architectures. Verdecchia et al. [15,16] employed a preliminary and a mixed-method case study to unravel the patterns of technical debt's evolution and the correlation between technical debt and the number of microservices. De Toledo et al. [4] conducted an exploratory case study to identify issues, solutions, and risks associated with the technical debt of microservices architecture. They [3] also conducted a multiple exploratory case study of three large companies to identify and prioritize the accumulation and prioritization of technical debt during microservices migration and propose a prioritization methodology. In addition, Bogner et al. [1] conducted an industry survey to collect information on how the industry prevents the accumulation of unknown technical debt related to maintainability, especially in microservices-based systems.

Some studies focused on different types of TD in microservices, such as bad smells [10], architectural TD [14], and migration TD [6]. However, the current TD studies mainly focus on the microservices architecture rather than cloud-native architecture, they are similar but different. In addition, existing research has focused primarily on patterns, types, and identification of TDs in microservices, with relatively few assessment and management strategies. The study aims to fill these research gaps, exploring the patterns, types, identification, evaluation and management strategies of TD in cloud-native applications to help development teams better address the challenges posed by TD and improve the quality and maintainability of applications.

3 Objectives

The objective of the research is to thoroughly investigate and analyze the impact of TD issues on software quality and maintenance in cloud-native applications. In order to achieve the stated objectives, the whole research process is divided into three concrete phases by answering the following research questions ($\mathbf{RQ}_s$):

- **RQ_1**. What are the common TD patterns in cloud-native applications? How do they affect the quality of the applications?
- **RQ_2**. How to identify and evaluate the TD in cloud-native applications?
- **RQ_3**. How to manage and mitigate the TD in cloud-native applications?

The theoretical basis for the whole research is built on the understanding of the concept of TD and cloud-native architectural applications. Many common TD patterns exist in cloud-native applications. These patterns may involve service coupling in microservice architectures, resource management in containerized deployments, and automated testing in serverless architectures. These TD patterns could lead to a decrease in quality and an increase in the maintenance costs of the application. Therefore, it is critical to understand these common TD patterns and their impact on application quality and maintenance.

How to identify and evaluate TD in cloud-native applications is also a key question. Identifying TD may involve methods such as code review, static code analysis, and metrics analysis. Evaluating the impact of TD may require considering its impact on software quality, maintainability, performance, etc. An in-depth analysis of the types, impact factors, and accumulation of TD can provide development teams with a more accurate method of identification and assessment, leading to a better understanding of the full picture of TD and the development of appropriate response strategies.

For existing TDs, how to better manage and mitigate them is another important area of research. Managing TD may involve prioritization, TD list management, team collaboration, and communication. Mitigating TD may require improving the quality and structure of existing code through refactoring, code restructuring, automated testing, and so on. In addition, regular TD cleanup and continuous TD monitoring are important tools for managing and mitigating TD. Therefore, investigating more effective ways of managing and mitigating TD and proposing constructive frameworks or tools can help development teams better cope with the challenges posed by TD and improve the quality and maintainability of applications.

4 Research Methods

At the beginning of the research, observation, literature review, and empirical study would be the primary research methodologies. A literature review of existing papers to understand the current state of research on TD under cloud-native applications, including the TD model and its impact on the quality and maintenance of the applications. A case study would also collect real industry data to gain insights into the performance and impact of TD in cloud-native applications. Qualitative research, code review, and data analysis allow for identifying and evaluating TD in cloud-native systems. Finally, exploratory research is used to research more effective methods, frameworks, or tools for managing and mitigating TD. Other types of empirical research will also be included if needed.

5 Working Plan and Expected Results

Our research is planned for three years. We plan to complete each research question over a year's time, and split each research question into 1–2 tasks and identify the research methods that could be required for the corresponding task. The specific working plan is presented in Table 1.

Table 1. Research Questions, Tasks, and Methods

Research Question	Tasks	Methods
1. What are the common TD patterns in cloud-native applications? How do they affect the quality and maintenance of the application?	1) Research on common TD patterns in cloud-native applications. 2) Research on the impact of TD patterns on quality and maintenance in cloud-native applications.	– Observation – Literature Review – Case Study
2. How to identify and evaluate the TDs in cloud-native applications?	1) Research on identifying the TDs in cloud-native applications. 2) Research on evaluating the TDs in cloud-native applications.	– Literature Review – Case Study – Data Analysis – Qualitative research – Code Review – Empirical Study
3. How to better manage and mitigate the TD that already exists in cloud-native applications?	1) Propose TD identification, evaluation, and management tools/frameworks/ methodologies for cloud-native applications based on the findings.	– Explorative Study – Case Study – Empirical Study – Data Analysis – Literature Review

Based on the research working plan, the research work is expected to be accomplished in the coming three academic years (Aug. 2024 - Jul. 2027). Therefore, the details of the working plan for the research and expected results are given in Table 2.

Table 2. Research Plan

Time	Themes	Target
Aug. 2024 – Jul. 2025	Literature review on technical patterns in cloud-native applications. Case study on impact of TD patterns on quality and maintenance in cloud-native applications.	RQ 1 (2 Publications)
Aug. 2025 – Jul. 2026	Quantitative analysis and qualitative research on identifying the TDs in cloud-native applications. Code Review, Empirical Study, and Data Analysis on evaluate the TDs in cloud-native applications.	RQ 2 (2 Publications)
Aug. 2026 – Jul. 2027	Explorative Study and Empirical Study on TD identification, evaluation, and management tools/frameworks/methodologies for cloud-native applications.	RQ 3 (2 Publications)
Summer 2027	Dissertation Preparation	-
Fall 2027	Dissertation Defense	-

Regarding our expected results (target) and potential publication venues, we plan to publish in conferences and journals related to Software Architecture and Software Quality and Maintenance, e.g., ICSA, ECSA, EASE, SANER, ESEM, etc. The evaluation of results depends on whether my research objectives are complete and papers are published successfully.

6 Reflection

Critical reflection on doctoral research work is essential to my academic growth and the advancement of knowledge in the field. In this research, we reflect on the main challenge that could be encountered, which is the dynamic nature of cloud-native architectures. The flexibility, extensibility, and continuous evolution of cloud-native architectures will naturally introduce cases of technical debt. As these applications are developed iteratively, the identification and management of technical debt will become increasingly complex. In addition, a deeper understanding of the differences in technical debt requires a comprehensive understanding of the patterns of its manifestation in cloud-native architectures. While existing literature provides insights into technical debt in the context of microservices software development paradigms, its translation to cloud-native environments has not yet been fully explored. Since my research is still in the beginning stages, these possible challenges need communication and collaboration with colleagues to gain insights, develop innovative solutions, and advance our understanding of technical debt in the context of cloud-native architectures.

7 Conclusion

The expected results of the research will include exploring common patterns of TD in cloud-native applications and their impact on quality and maintenance. Also, this inquiry will research methods or frameworks for identifying and evaluating TD in cloud-native applications. In addition, we will propose a strategy or approach to manage and mitigate TD in cloud-native applications. The research results are supposed to be published in international and regional journals and conferences in relevant areas.

Disclosure of Interests. The author has no competing interests to declare that are relevant to the content of this article.

References

1. Bogner, J., Fritzsch, J., Wagner, S., Zimmermann, A.: Limiting technical debt with maintainability assurance: an industry survey on used techniques and differences with service-and microservice-based systems. In: Proceedings of the 2018 International Conference on Technical Debt, pp. 125–133 (2018)
2. Cerny, T., Taibi, D.: Static analysis tools in the era of cloud-native systems. arXiv preprint arXiv:2205.08527 (2022)
3. De Toledo, S.S., Martini, A., Nguyen, P.H., Sjøberg, D.I.: Accumulation and prioritization of architectural debt in three companies migrating to microservices. IEEE Access **10**, 37422–37445 (2022)
4. De Toledo, S.S., Martini, A., Przybyszewska, A., Sjøberg, D.I.: Architectural technical debt in microservices: a case study in a large company. In: 2019 IEEE/ACM International Conference on Technical Debt (TechDebt), pp. 78–87. IEEE (2019)
5. Kratzke, N., Siegfried, R.: Towards cloud-native simulations-lessons learned from the front-line of cloud computing. J. Defense Model. Simul. **18**(1), 39–58 (2021)
6. Lenarduzzi, V., Lomio, F., Saarimäki, N., Taibi, D.: Does migrating a monolithic system to microservices decrease the technical debt? J. Syst. Softw. **169**, 110710 (2020)
7. Lomio, F., Moreschini, S., Li, X., Lenarduzzi, V.: Anomaly detection in cloud-native systems. In: 2022 48th Euromicro Conference on Software Engineering and Advanced Applications (SEAA), pp. 100–103. IEEE (2022)
8. Long, X., Gang, L., Qi, Z., Huiyan, Z., Tingjun, W.: Cloud native intelligent operation and maintenance technology. Telecommun. Sci. **36**(12) (2020)
9. Moreschini, S., Younesian, E., Hästbacka, D., Albano, M., Hošek, J., Taibi, D.: Edge to cloud tools: a multivocal literature review. J. Syst. Softw. 111942 (2023)
10. Pigazzini, I., Fontana, F.A., Lenarduzzi, V., Taibi, D.: Towards microservice smells detection. In: Proceedings of the 3rd International Conference on Technical Debt, pp. 92–97 (2020)
11. Rantala, L., Mäntylä, M., Lenarduzzi, V.: Keyword-labeled self-admitted technical debt and static code analysis have significant relationship but limited overlap. Softw. Qual. J. 1–39 (2023)

12. Robredo, M., Saarimäki, N., Peñaloza, R., Taibi, D., Lenarduzzi, V.: A comparison between multivariate time series analysis and machine learning techniques for technical debt prediction (2024)
13. Tapia, V., Gaona, M.: Research opportunities in microservices quality assessment: a systematic literature review. J. Adv. Inf. Technol. **14**(5) (2023)
14. de Toledo, S.S., Martini, A., Sjøberg, D.I., Przybyszewska, A., Frandsen, J.S.: Reducing incidents in microservices by repaying architectural technical debt. In: 2021 47th Euromicro Conference on Software Engineering and Advanced Applications (SEAA), pp. 196–205. IEEE (2021)
15. Verdecchia, R., Maggi, K.: Technical debt in microservices: a mixed-method case study.
16. Verdecchia, R., Maggi, K., Scommegna, L., Vicario, E.: Tracing the footsteps of technical debt in microservices: a preliminary case study. In: International Workshop on Quality in Software Architecture (2023)
17. Villa, A., Ocharán-Hernández, J.O., Pérez-Arriaga, J.C., Limón, X.: A systematic mapping study on technical debt in microservices. In: 2022 10th International Conference in Software Engineering Research and Innovation (CONISOFT), pp. 182–191. IEEE (2022)
18. Waseem, M., Liang, P., Shahin, M., Ahmad, A., Nassab, A.R.: On the nature of issues in five open source microservices systems: an empirical study. In: Proceedings of the 25th International Conference on Evaluation and Assessment in Software Engineering, pp. 201–210 (2021)

Improving QoS of Microservices Architecture Using Machine Learning Techniques

Neha Kaushik[✉]

J.C, Bose University of Science and Technology, Faridabad, India
20001902008@jcboseust.ac.in

Abstract. Microservices architecture has gained significant popularity in the development of modern software applications due to its scalability, flexibility, and modularity. However, ensuring high-quality service delivery while maintaining the agility and responsiveness of microservices poses several challenges. This paper introduces an innovative method aimed at enhancing the Quality of Service (QoS) in microservices architecture-driven applications through the utilization of machine learning techniques. Initially, the primary factors contributing to the overall quality of microservices applications are identified. Subsequently, a machine learning-based framework is proposed for enhancing the QoS of such applications. To validate this framework, experimental assessments are conducted using sample microservices applications as case studies. The outcomes of these experiments demonstrate a significant enhancement in the overall QoS of the microservices application facilitated by the proposed framework.

Keywords: Microservices architecture (MSA) · Quality of Service (QoS) · Performance · Reliability

1 Introduction

Microservice Architectures (MSA) emerged in the context of developing applications as a suite of highly cohesive and finely grained services. The concept of microservices architecture was firstly discussed by James Lewis and Martin Fowler. Each microservice is responsible for performing only a single task representing a business capability and uses lightweight mechanisms such as HTTP APIs for communicating with other microservices [1]. MSA enables each microservice to be developed, tested, deployed, and upgraded independently [2].

Microservices architecture can be seen as the extension of Service Oriented Architectures (SOA). There are some similarities in microservices and SOA based systems like in both architectures complex applications are decomposed into the set of loosely coupled services. SOA based applications become monolithic as the size of the application increases and this makes the applications complex and difficult to manage [4]. Large organizations such as Amazon, Netflix, Uber, Spotify, Twitter, and Google have started developing their applications using microservices architecture as a result of numerous advantages. Despite several benefits, microservices is still a relatively new technology

© The Author(s), under exclusive license to Springer Nature Switzerland AG 2024
A. Ampatzoglou et al. (Eds.): ECSA 2024, LNCS 14937, pp. 72–79, 2024.
https://doi.org/10.1007/978-3-031-71246-3_9

and needs to explore more [3]. Quality assurance, data consistency and debugging are the challenges that need to be addressed [7]. [5, 6] have considered quality assurance as a key concern during the migration or the development of microservices architecture based applications. To the best of our knowledge prior research has looked into some of the technical difficulties related to the implementation of microservices, assessing the quality of already developed microservices application but there is a lack of work done in order to improve the quality of microservices application. In this paper, an attempt is made to fill the gap by identifying the quality attributes contributing the most to the quality of microservices application and then proposing a framework for improving the quality based on the identified quality attributes.

The main contribution of this paper can be summarized as:

1. Identification of the quality attributes contributing the most in the overall QoS of microservices architecture based applications.
2. Development of a model for improving the performance, reliability and maintainability of the microservices applications respectively.
3. An empirical analysis to compare the performance, reliability and maintainability of case study microservices applications with and without proposed framework.
4. A machine learning based model to predict the response time of microservices architecture based applications which can be used for making performance improvement decisions. And also, Machine learning has been used to take a proactive action rather than a reactive action for reliability improvement.

Paper Structure: Sect. 2 demonstrates the relevance of the work. In Sect. 3, we provide the objectives of the thesis. Section 4 describes the thesis in a nutshell. Section 5 concludes the work and provides some potential future directions from the thesis. In Sect. 4, we provide the progress and outlook of the thesis.

2 Relevance of the Work

Microservices is a relatively new architectural style widely adopted by large organisations such as Amazon, Netflix, LinkedIn, SoundCloud, and many others for application development [8]. Because microservices is a new architectural style with several benefits for application development there is a need for a broad and systematic analysis of the tactics to improve specific quality attributes contributing the most in the overall quality of the microservices application. Due to the novelty of microservices as an architectural style, which offers numerous advantages for application development, it is essential to conduct a systematic analysis of strategies aimed at enhancing the specific quality attributes that have the greatest impact on the overall quality of microservices applications [6]. Wang et al. [5] and Li et al. [6] have found out that performance, reliability and security are the quality attributes that contribute the most in the overall quality of MSA based application and also suggested working in these areas for their improvement.

Abdullah et al. [9] proposed a burst aware autoscaling method to solve the problem of detecting and handling burst workload in containerized microservices based applications in order to preserve performance. The proposed model resulted in performance improvement but the limitation of the proposed model is that it was tested on a single microservice but not on the application containing multiple microservices. Joseph and Chandrasekaran [10] reduced the response time of microservices and performed efficient utilisation of resources by deploying the microservices based on their interaction with other microservices. However the author focused on the efficient utilisation of resources but not considered the impact of network characteristics such as link congestion before making deployment decisions. Additionally, energy efficiency of nodes is also a key point to consider in deployment decisions, which was ignored in the proposed model of the author. Sampaio Jr. et al. [11] has used a three step approach for optimising the performance of microservices application. The approach used by the author sometimes improves the performance but at the same time in some cases it is also responsible for performance degradation. Camilli and Russo [12] focused on identifying the microservices which can lead to performance degradation of complete applications. The results of the proposed model can be used by the engineers during system maintenance. Cortellessa et al. [13] monitors the runtime interactions of all microservices of an application and then uses the monitored data for improving the performance by re-engineering the design of the application. In the proposed approach, the refactoring decisions are fully automated, and it is not a good practice to neglect human expertise in software re-engineering. Based on the analysis of the existing work for performance improvement of microservices applications, it is found that most of the researchers have been focusing on the performance improvement of microservices application by working on the operational stage. None of the existing work has focused on performance improvement by working in the development phase of the application. Making improvements on the early stage is cost effective and easier to do.

Liu et al. [14] has considered reliability improvement as an important parameter for improving the quality of cloud based microservices applications. The author firstly used predicted petri nets for modelling the components of microservices application, then scheduled the microservices in such a way that dependent microservices were placed together resulting in improved reliability of the application. As future work, the author suggested including environment diversity for predicting the reliability of application. Song and Tilevich [15] have proposed a data flow based programming model that can be used to develop real time microservices applications. The proposed model results in better resource utilization and improved reliability. The author has focused on the development phase reliability improvement while ignoring operational reliability of the application. Pietrantuono et al. [16] have proposed EMART, an operational reliability assessment method of microservices application. The proposed method used failing/success of user requests and data about microservices usage for reliability estimation of the application. The author ignored service interaction information, which affects the overall reliability of the application. The author suggested using the resulting data in reliability improvement of the applications. Malki et al. [17] have considered the workload configurations and predefined rate limits for predicting their effect on the reliability of the microservices, the author found that limiting the API rate significantly improves the reliability of microservices applications but also impacts the performance and scalability of the application. Jagadeesan and Mendiratta [18] have proposed a service mesh and side cars based framework to detect and handle the microservices failures that can impact the reliability of application. Ding et al. [19] have emphasized to consider the runtime environment and software dynamic behaviour for reliability prediction and reliability improvement of Service oriented Systems. The author has proposed a framework which uses run time data such as response time to predict and improve the reliability of Service Oriented Systems. The literature on reliability improvement shows that the reliability attribute contributes the most in the overall quality of the microservices application and a lack of work has been done for improving the reliability of the application. It is also observed that parameters such as response time, throughput can be used for the reliability improvement of an application. This study helped in understanding the importance of proactive action in comparison to reactive action. The literature review motivated to propose a framework focusing on the reliability improvement of microservices application using a proactive approach.

3 Thesis Statement

The empirical data presented in Sec. 2 has guided us in formulating the thesis statement, which is outlined below.

Work 1: Performance improvement and prediction for microservices architecture based applications using machine learning.

Work 2: To predict and improve the reliability of microservices architecture based applications using machine learning algorithms.

Work 3: To predict and improve the maintainability of microservices architecture based applications using machine learning algorithms.

4 Current State of the Thesis

In the following we present the main research activities of the thesis. Predicting and improving the maintainability of microservices architecture based applications is not discussed in detailed, as this work is still under progress.

4.1 Performance Improvement and Prediction for Microservices Architecture Based Applications Using Machine Learning

Motivation. The performance of MSA based application can be measured in terms of the time it takes to respond to user requests i.e. Response Time, the number of requests completed per unit time i.e. Throughput or the number of resources an application uses to serve the user request i.e. Resource usage. The lifecycle of a microservices application can be divided into two phases, developmental phase and the operational phase [11]. The performance of an application can be improved by working on both phases of the application. But to the best of our knowledge, several researchers have suggested improving the performance by focusing on the granularity of the microservice but in actuality there is a lack of work done in order to optimize the performance of a microservice architecture based application by focusing on its developmental phase [20–22]. There is still a scope where by making changes to the architecture of microservices architecture based web applications, their performance can be improved. Moreover, making changes during the early stage of the application development not only improves the performance of the web application but it will also be less complex and cost effective.

Approach. In the proposed model, micro frontend following microservices in the backend is used. The application is designed in such a way that the whole application will be divided into small and independent services. Each service will be responsible for performing a single task and will have access to their separate database. Every webpage of the application can consist of a single or multiple services depending upon its requirement. A single team will be responsible for the designing the frontend, backend and maintenance of the service. All the services can be developed using different programming languages. When using micro frontends and microservices together, the front-end modules are developed and deployed independently, but they still need to communicate with the back-end services. This can be done using well-defined APIs and a gateway that handles the communication between the front-end modules and the back-end services. The division of whole application into individual services make it easier to design and develop it, allow faster delivery of the application to the client, loosely coupled services require less communication hence reduces communication overhead and improves response time, making changes to one application do not affects the other services and let other services work hence it is easier to maintain the application developed using proposed model, independent services makes the deployment and scaling operations easy to implement and cost effective.

Findings. To perform the empirical analysis, we have developed an Eshop application using monolithic frontend followed by microservices in the backend and micro frontend followed by microservices in the backend. Observing the performance matrix of all the microservices and the graphical representation of response time and throughput, we can conclude that the proposed model is giving better results than the existing model. The comparison between both the approaches is based on the response time, error percentage and throughput matrix. And in all the three matrix frontend based applications lead to reduced response time i.e. faster response to the end user for their http requests, increased the throughput and reduced the number of failed requests. So seeing the results we can conclude that the proposed model is more efficient (in terms of performance) than the monolithic frontend followed by microservices in the backend.

4.2 Predicting and Improving the Reliability of Microservices Architecture Based Applications Using Machine Learning

Motivation. According to the IEEE standard, Reliability is characterized as the capacity of an application to execute its designated functions within specified conditions throughout a defined duration [23]. Microservices applications become unreliable or fail to respond to the user in specific time in the presence of bursty workloads [9]. The services in a MSA based application are loosely coupled i.e. they rely on other services for some functionalities. So, when a microservice experiences the bursty workload it can have a cascading effect on its dependent services. This can lead to a breakdown in communication between services and affect the overall reliability of the application [18].

If the application is not properly prepared for handling the bursty workload then it can lead to resource exhaustion and can make the application unresponsive. Microservices architectures need to be resilient and capable of recovering from such failures because if failures occur and the system is unable to recover quickly, it can lead to prolonged periods of reduced reliability.

Approach. The proposed framework includes four phases. Phase 1 is the Reliability Assessment of microservices application. In this phase the default reliability of the application will be assessed against the bursty workload. In phase 2, a machine learning model has been trained for predicting the number of instances required for handling the user requests successfully. Phase 3 of the proposed framework focuses on the prediction of the bursty workload and uses the model trained in phase 2 to predict the number of microservice instances required for handling that workload. This way the application becomes reliable for handling the dynamic environment and the user requests are serviced without a failure. In the last phase of the proposed framework tha task is to place the predicted instances on the nodes such that the microservices that communicate more frequently are placed on the same node. Communication aware placement of microservices reduces the network communication overhead and leads to the reduced response time of the application.

Findings. A microservices application named Microservice Consul for evaluating and comparing the performance of proposed framework. After analysing the performance of individual microservices and complete application, it can be concluded that the proposed framework is improving the number of successfully processed user requests and reducing the number of failed requests. This reduction of failed requests in user requests results in improved reliability of the complete application. So we can say that the proposed framework is making the microservice architecture based application more reliable for dynamic environments like bursty workloads.

5 Conclusions, Contributions, and Future Directions

The objectives of the thesis are divided into three parts. The first objective focuses on the performance improvement of MSA based applications and to use machine learning for predicting the performance of microservices applications. For this purpose a framework including Micro Frontend followed by Microservices in the backend has been proposed and results show that it significantly increased the performance of MSA based applications. The second objective focused on improving the reliability of microservices applications. For that purpose a framework is proposed for making the MSA based application reliable for bursty workload situations. The experimental results shows that the presence of the proposed framework makes the microservices application more reliable for bursty workload situations. The third objective is to improve the maintainability of the microservices applications. MSA based applications are difficult to maintain because of their distributed nature. So the third objective focuses on the maintainability improvement of microservices applications. This objective is still in progress. Through the literature survey and experimental analysis of proposed frameworks we conclude that:

1) Performance, reliability, and maintainability are the quality attributes that contribute the most in the overall QoS of microservices architecture based applications.
2) Micro Frontend followed by microservices improves the performance of microservices architecture based applications significantly.
3) Response time and throughput are the attributes that can be used to improve the reliability of microservices application to make it reliable for handling bursty workload situations.

In future, the performance prediction using the response time of microservices application can be used to improve the performance of microservices applications in the operational phase.

6 Progress and Outlook

The paper titled "Micro Frontend Based Performance Improvement and Prediction for Microservices Using Machine Learning" from work 1 has been published in the Journal of Grid Computing in the year 2024, and "Empirical Evaluation of Microservices Architecture" has been published in the Communication and Intelligent Systems: Proceedings of ICCIS conference, 2022. The author is working on writing work 2 and planning to submit it in the Journal of Software: Practice and Experience. The author is expecting to defend her thesis in the beginning of 2025.

Acknowledgments. The author thanks her Ph.D/Thesis supervisors Dr. Vinay Raj and Dr. Harish Kumar for the help in completing the PhD dissertation.

References

1. Raj, V., Sadam, R.: Evaluation of SOA-based web services and microservices architecture using complexity metrics. SN Comput. Sci. **2**(5), 1–10 (2021)
2. Taibi, D., Spillner, J., Wawruch, K.: Serverless computing-where are we now, and where are we heading? IEEE Softw. **38**(1), 25–31 (2020)
3. Singh, A., Raj, V., Ravichandra, S.: Integration of attribute-based access control in microservices architecture. In: ICT Systems and Sustainability, pp. 681–690 Springer, Singapore (2022). https://doi.org/10.1007/978-981-16-5987-4_69
4. Dragoni, N., et al.: Yesterday, today, and tomorrow. Present Ulterior Softw. Eng. 195–216 (2017)
5. Wang, Y., Kadiyala, H., Rubin, J.: Promises and challenges of microservices: an exploratory study. Empirical Softw. Eng. **26**(4), 63 (2021)
6. Li, S., et al.: Understanding and addressing quality attributes of microservices architecture: a systematic literature review. Inf. Softw. Technol. **131**, 106449 (2021).
7. Ghofrani, J., Lübke, D.: Challenges of microservices architecture: a Survey on the State of the Practice. ZEUS, 1-8 (2018)

8. Taibi, D., Lenarduzzi, V., Pahl, C.: Processes, motivations, and issues for migrating to microservices architectures: an empirical investigation. IEEE Cloud Comput. **4**(5), 22–32 (2017)
9. Abdullah, M., Iqbal, W., Berral, J.L., Polo, J., Carrera, D.: Burst-aware predictive autoscaling for containerized microservices. IEEE Trans. Serv. Comput. **15**(3), 1448–1460 (2020)
10. Joseph, C.T., Chandrasekaran, K.: IntMA: Dynamic interaction-aware resource allocation for containerized microservices in cloud environments. J. Syst. Archit. **111**, 101785 (2020)
11. Sampaio, A.R., Rubin, J., Beschastnikh, I., Rosa, N.S.: Improving microservice-based applications with runtime placement adaptation. J. Int. Serv. Appl. **10**(1), 1–30 (2019)
12. Camilli, M., Russo, B.: Modeling performance of microservices systems with growth theory. Empirical Softw. Eng. **27**(2), 39 (2022)
13. Cortellessa, V., Di Pompeo, D., Eramo, R., Tucci, M.: A model-driven approach for continuous performance engineering in microservice-based systems. J. Syst. Softw. **183**, 111084 (2022)
14. Liu, Z., Yu, H., Fan, G., Chen, L.: Reliability modelling and optimization for microservice-based cloud application using multi-agent system. IET Commun. **16**(10), 1182–1199 (2022)
15. Song, Z., Tilevich, E.: Equivalence-enhanced microservice workflow orchestration to efficiently increase reliability. In 2019 IEEE International Conference on Web Services (ICWS), pp. 426–433. IEEE (2019)
16. Pietrantuono, R., Russo, S., Guerriero, A.: Testing microservice architectures for operational reliability. Softw. Test. Verification Reliab. **30**(2), e1725 (2020)
17. El Malki, A., Zdun, U., Pautasso, C.: Impact of API rate limit on reliability of microservices-based architectures. In 2022 IEEE International Conference on Service-Oriented System Engineering (SOSE), pp. 19–28. IEEE (2022)
18. Jagadeesan, L. J., & Mendiratta, V. B.: When failure is (not) an option: reliability models for microservices architectures. In 2020 IEEE International Symposium on Software Reliability Engineering Workshops (ISSREW), pp. 19–24. IEEE (2020)
19. Ding, Z., Xu, T., Ye, T., Zhou, Y.: Online prediction and improvement of reliability for service oriented systems. IEEE Trans. Reliab. **65**(3), 1133–1148 (2015)
20. Al Qassem, L.M., Stouraitis, T., Damiani, E., Elfadel, I.A.M.: Proactive random-forest autoscaler for microservice resource allocation. IEEE Access **11** (2023)
21. Yan, M., Liang, X., Lu, Z., Wu, J., Zhang, W.: HANSEL: Adaptive horizontal scaling of microservices using Bi-LSTM. Appl. Soft Comput. **105**, 107216 (2021)
22. Xu, M., CoScal: Multi-faceted scaling of microservices with reinforcement learning. IEEE Trans. Netw. Serv. Manage. (2022).
23. Wang, H., Wang, L., Yu, Q., Zheng, Z., Bouguettaya, A., Lyu, M.R.: Online reliability prediction via motifs-based dynamic Bayesian networks for service-oriented systems. IEEE Trans. Soft. Eng. **43**(6), 556–579 (2016)

CASA Workshop

Designing, Implementing, and Testing AI-Oriented Smart Home Applications: Challenges and Best Practices

Denivan Campos[✉], Luana Martins[✉], Joselito Mota[✉], Dhyego Tavares[✉], Jander Pereira[✉], Mayki Oliveira[✉], Denis Boaventura[✉], Diego Correa[✉], Eduardo Ferreira[✉], George Pinto[✉], Nilton Seixas[✉], Adriano Maia[✉], Matias Romário[✉], Ernando Passos[✉], Frederico Durao[✉], Gustavo B. Figueiredo[✉], Maycon Peixoto[✉], Tiago Januario[✉], Cassio Prazeres[✉], Ivan Machado[✉], and Eduardo Almeida[✉]

Federal University of Bahia - UFBA, Salvador, BA, Brazil
{denivan.campos,martins.luana,joselito.mota,dhyegocruz,
jandersantos,maykioliveira,denis.boaventura,diego.correa,
eduardoferreira,george.pacheco,nilton.seixas,adriano.maia,matiasrps,
ernando.passos,fdurao,gustavobf,maycon.leone,januario,prazeres,
ivan.machado,eduardo.almeida}@ufba.br

Abstract. Smart homes represent lifestyles that enhance the daily lives of homeowners compared to non-smart homes, facilitating and improving the quality of life of the general residents. While home automation is already a reality, ongoing studies are being conducted to integrate smart scenes and rules. This paper presents an exploratory study on smart home development in a large Brazilian software company. Through semi-structured interviews, we investigate the challenges practitioners face in developing smart homes. In addition, we propose a set of best practices based on the strategies practitioners adopt to overcome the challenges.

Keywords: Artificial Intelligence · Smart Home · Challenges and Strategies

1 Introduction

In the last few decades, smart homes have become the subject of active research into intelligent environments due to advances in engineering and technology [7]. Smart homes represent lifestyles that enhance the daily lives of homeowners compared to non-smart homes [3]. According to the Smart Homes Association, a smart home is defined as a technological environment that aims to facilitate and improve the quality of life of the general residents in carrying out their daily living activities, thus increasing their autonomy in a single location [5,8].

Smart homes integrate intelligent technology such as numerous sensors and connected devices that can communicate with each other and be controlled via hub or smartphone to ensure greater control, security, comfort, and automation in homes [9]. Users can remotely control and monitor smart home equipment using the telecommunications-based remote monitoring system [6].

Home automation is a reality. Fernandes et al. [4] and Sikder et al. [9] show that there has been an evolution in devices' capabilities, which have evolved from just controlling lights and opening garage doors to connecting entire residential spaces, enabling more autonomous and efficient daily operations. Generally, smart home users manually define trigger-action rules to configure scenarios using sensors and devices. The sensors provide information to the devices, which can make automated decisions based on this input [9]. For example, in a trigger-action smart home, the system relies on simple and rigid rules such as "Turn on the kitchen light at 6 a.m. when motion is detected."

The decision-making process of device actions should lean towards Artificial Intelligence (AI) rather than human intervention to create a completely autonomous smart home [9]. An AI-oriented smart home dynamically adapts to residents' behavior patterns. The AI learns from interactions and habits, allowing continuous, context-sensitive personalization. For instance, the AI might notice a resident wakes up at 4 a.m. on workdays but has a different weekend routine, automatically adjusting the home's settings to match these nuanced preferences. However, smart homes with such a level of automation are still rare.

A collaboration between a large company (from now named COMPANY A) and practitioners from the Applied Artificial Intelligence Laboratory (LIAA) implemented an AI-oriented smart home that utilizes AI to suggest actions for devices based on inferred user living habits, thereby creating an orchestrated and intelligent behavior for smart home users. An example of such an intelligent behavior recommendation is when opening the door, and the system recommends plugging in the external light. When closing the door, it suggests turning off the external light. When entering the room and opening the window, the system understands that cleaning should begin and triggers a robot vacuum cleaner. When the window is closed, the vacuum robot finishes the cleaning process.

To solve the above problem, the practitioners first carried out an ad hoc literature review, examining the state of the art in each of the project's architectural areas (see Smart Homes Architecture section). Nevertheless, there is a need for more evidence in the literature on the difficulties practitioners face and the best practices they employ in constructing smart homes focused on using AI without human intervention. This paper aims to identify the challenges practitioners face and the practices they adopt when developing a smart home. As an outcome, the paper presents guidelines to overcome the identified challenges and good practices ranging from design, implementation, and testing of smart homes. We conducted semi-structured interviews with the team of practitioners and collected a set of data to answer the following research questions (RQs):

RQ1. What are the challenges practitioners face in developing asmart home solutions?

RQ2. What are the best practices adopted by practitioners for developing a smart home solution?

Our study has two main contributions. First, it identifies practitioners' main challenges in developing smart homes and the essential practices adopted. Next, it proposes a set of guidelines to solve these identified challenges.

2 Smart Homes Architecture

This section introduces our AI-oriented smart home architecture. Figure 1 shows the modules and essential elements of the smart home architecture. Each module is color-coded to indicate their specific origins within our collaboration. The green and yellow boxes emphasize the A company's and third-party collaborators' contributions, respectively. On the other hand, the purple boxes symbolize the modules developed by the practitioners, representing an innovative, customized solution for Smart Homes. Next, we present each of the modules.

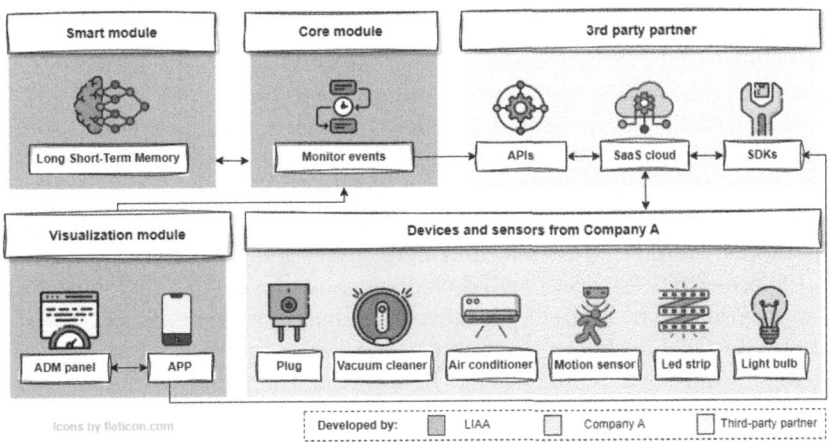

Fig. 1. Smart Home Architecture

Smart System: The intelligent system consists of two modules: *Core* and *Smart* modules developed by the practitioners team. The *Core* module is responsible for acquiring data via endpoints from third-party partner Application Programming Interface (APIs) and monitoring events. It also fetches the status of devices and provides information for the *Smart* Module. The *Smart* module, in turn, features an AI based on *Long Short-Term Memory (LSTM)* that learns the following steps from the user's recurring data to recommend actions within the home. All recommendations are shown to the user via a mobile app.

Visualization (APP): All user actions are performed through the app. The app uses third-party partner Software Development Kit (SDKs) to allow users to provide feedback for a recommendation and to manipulate the devices and homes. The APP is responsible for connecting devices, smart home residents, and the *Core* and *Smart* modules, which listen to actions and provide suggestions for new home recommendations. Besides providing a visualization for customers, the system administrator can access an interactive dashboard to visualize energy consumption, systems' accuracy, and device-related data.

Third-Party Partner: The third-party partner provides a software automation platform offering Internet of Things (IoT) solutions and connects these smart devices to the Software as a Service (SaaS) cloud and other devices, allowing the integration and control of the smart home ecosystem. In addition, the third-party partner provides support via APIs and SDKs, facilitating the integration and extensibility of the Smart Home Ecosystem.

Devices: The COMPANY A integrates sensors and other devices that form the backbone of smart homes, including motion sensors, presence sensors, light bulbs, switches, locks, sockets, and robot vacuum cleaners. The devices are the main interconnected objects in the IoT network.

3 Related Work

This section presents related works divided into two categories: overviews of the literature and practical applications in smart home technology.

Carra and Tabia [2] presented a concise overview of Smart Home solutions and a list of opportunities and challenges for AI. The solutions focus on smart home functionalities such as detecting risk situations and virtual assistants. The challenges and opportunities mainly focus on technical aspects of data security, interoperability, industrialization, and personalization. In another study, Singh et al. [10] presented an overview of challenges to developing IoT-based smart home applications. The main challenges are human motion detection, scalability, lack of standards, interoperability, data processing and management, mobility management, and device connectivity. In addition, the paper points out research directions based on the challenges and current applications of smart homes.

The studies performed by Carra et al. [2] and Singh et al. [10] investigated the challenges of developing smart homes from a theoretical view. In contrast, our study presents an exploratory investigation with semi-structured interviews to gather empirical data directly from industry practitioners. While the cited studies highlight challenges and future research directions, our study offers practical insights and actionable best practices derived from practitioners' experiences. This includes specific strategies they employ to overcome daily challenges in smart home development.

Focusing on smart home applications, Zhu et al. [14] proposed a design scheme by introducing a wireless sensor system and artificial speech recognition technology to ease the operation and improve the efficiency of information interaction. In the context of smart home surveillance, Thakur et al. [12] proposed a cost-effective, lightweight Edge AI-enabled IoT Framework using a Raspberry Pi and the open-source software Motion. The system monitors video signals from various cameras and notifies the homeowner via email and smartphone messages.

The studies performed by Zhu et al. [14] and Thakur et al. [12] are examples of recent applications of AI in smart homes. While they focus on a specific context, i.e., home surveillance and user interaction, the smart home presented in this study is applied to diverse contexts. In addition, we investigate the practitioner's perspectives regarding smart home development.

4 Research Design

The *goal* of the exploratory study is to investigate the challenges faced by practitioners in developing smart homes within a large Brazilian software company, with the *purpose* of understanding the strategies used to overcome these challenges. The *perspective* is of both researchers and practitioners who are interested in enhancing smart home technology and improving development practices to better integrate automation and AI-driven solutions into smart environments.

For the data collection, we conducted semi-structured online interviews with the practitioners' teams. First, we interviewed one of the team leaders to understand the architecture of the smart home, the division of activities, and work teams. In short, the team is composed of 15 practitioners divided into four specialized teams: (1) *Smart System:* composed of 3 practitioners in recommendation systems; (2) *Visualization (APP)* composed of 3 practitioners with expertise in architecture and software reuse for highly configurable systems; (3) *Third-Party Partner* composed of 6 practitioners with expertise in hardware, communication protocols, and IoT device modeling; and (4) *Devices* composed of 3 practitioners in system deployment (DevOps).

We grouped the interviewees based on the module they worked on in the Smart Home Architecture (Sect. 2). The Smart System practitioners worked on two modules (i.e., *Smart module* and *Core module*), resulting in a total of 5 groups. Each interview lasted an average of 30 to 40 min. The interview script is available in our online repository [1]. However, given a nondisclosure agreement (NDA) and confidentiality requirements, all the transcripts of the interviews will remain private.

We began the interviews with a brief introduction to the research objectives. Then, we asked them six questions on two main themes: (i) the main challenges faced during the development of the smart home and (ii) the solutions proposed to overcome these challenges.

After collecting the data, the authors transcribed the audio for the data analysis. We used the open code method as detailed by Stol et al. [11], which involves breaking down the interview transcripts into discrete parts, comparing and grouping them for similarities. Subsequently, we promoted discussions about coding in weekly practice meetings among the authors, aiming to improve the results' reliability and mitigate possible biases. Challenges related to the devices, the third-party partners, the visualization (app), the intelligent system, and DevOps were identified during this process. Then, we elaborated on the guidelines based on the solutions adopted by practitioners.

5 Challenges & Best Practices

5.1 *Guidelines for Developing Smart Homes*

We present the challenges and best practices discussed in each of the classifications covered in the research design section. These guidelines are directly related to the design of a specific project within a data context. For example, depending on the third-party partners, the form and variables of data collection may vary. However, the challenges and good practices identified can be applied generically to the development of any smart home based on IoT and AI, on the assumption that the data can be altered due to the way it is sent and received. In this context, we classified the challenges into four parts: Third-party Partner, Visualization (APP), Smart Systems, and DevOps, with Smart Systems being subdivided into Smart Module and Core.

CORE (CO): We present the challenges and good practices reported by the Core Module practitioners. We identified 19 challenges and 28 good practices for addressing them.

Challenge CO1. Distributed system saving memory and processor without adding network delay. The system was developed to work in a distributed way to isolate processes in different nodes. It helps the system have specialized code for each node/module. The core is the main module that administrates, delivers, and receives all data produced by all other modules. Based on it, the core module is also distributed, sharing the main database. It allows each part of the core to be in an isolated container, working on a single task. It helps to save processor and memory because each container works on a specialized task, which expands the time to finish.
Best Practice CO1. Practitioners use their experiences and search for industry-academia experiences over real-time systems to design the system.

Practitioners read papers about distributed and monolithic systems, seeking to observe the short and long-term advantages and disadvantages.

Challenge CO2. The algorithms' processing time might increase with the potential growth of devices being used in the smart home. Practitioners develop machine learning models that are not always in memory, and we also select algorithms based on time and precision.
Best Practice CO2. Implement bit counting. The bit count contributes to processing efficiency, and the data size count makes it possible to evaluate space allocation savings. Practitioners observe data repetition and code variables in memory. Practitioners preprocess the dataset to reduce the amount of data in memory and load code in memory to process the data.

Challenge CO3. When sharing data between modules, the network delays approximately 20 s. The smart devices produce continuous data to the core. The smart module produces data when there are recommendations. The app produces data when users interact with the devices or recommendations.

Best Practice CO3. During implementation, it is essential to check for network delay. Network delay refers to the time the transmitted data is delayed between devices on a network. Various factors, such as network bandwidth, traffic congestion, physical distance between devices, and processing time, can cause this delay. Typically, network delay can cause poor communication performance and the response time of applications and online services. Evaluating the type of connections between core and other modules, varying which one is a single connection or a continuous connection. It was also observed the amount of data between the core and other modules to decide if the connection will be stated by the core or by the other modules.

Challenge CO4. Establish effective communication during module integration. The third-party connections are in Apache Listener, or Message Queuing Telemetry Transport (MQTT), in the way of listening to the devices. The connection with our modules is through Hypertext Transfer Protocol (HTTP). The connection to send device commands is in Socket.

Best Practice CO4. Study the code's functionality and have professional experts' support. Code development considers all these connections, isolating each type of connection so as not to interfere with the system's performance.

Challenge CO5. Scheduling background activities to be triggered daily or every seven days, or when processors are operating at 100%, ensuring that they do not cause crashes or overload the machine. This involves efficiently distributing the background processing load via COM or Celery. Verification of the amount of energy consumed in a home. The system precision is obtained by observing the user's feedback during the day and week.

Best Practice CO5. Training all homes once a week; for example, `Spotify` generates music for everyone once a week, whereas, on Monday, there are new songs on the list to listen to. In the case of Smart Home, this cannot happen. So, we modeled it so that each apprenticeship in each home would have its specific training day. It is based on when the home enters the system, i.e., when it registers so that it does not occupy the processors. Isolating this job in specialized containers that run its verification time in time.

Challenge CO6. The practitioners have an instability in the system because each device is published differently. They cannot load the different data in a consulted class. A RGB light version 1 and 2 publishes different types of messages. However, to the users, it is the same light.

Best Practice CO6. Use a third-party partner platform. It could earn an average of six months' work, especially if the team is not very experienced or has little previous experience. Modeling the database to accept this difference and process it in an isolated way, showing it to the user in the same pattern.

Challenge CO7. To create a code environment that allows local development, lab tests, whole system validation, and a production system.
Best Practice CO7. Use a system of configuration variables loaded at startup depending on the environment it is running. We can make various combinations such as running everything on `Localhost`, running everything on the same machine, running everything distributed in the lab, running everything locally inside the docker, and running everything distributed inside the docker on the development servers. The code must be adapted to software implementation, which varies depending on the loaded environment. Where the IP address self-adapts to the environment parameters.

Challenge CO8. Synchronization itself, along with multiprocessing and distributed multiprocessing. To allow the modules to be synchronized, receiving data in real-time.
Best Practice CO8. Setting variables at the beginning and implementing multiprocessing points in the system allows the distributed system to multi-process the data without job dependency, keeping the system synchronization.

Challenge CO9.1. Implementation so that other people can read and understand it easily, making the system available for production.
Best Practice CO9.1. The code is designed to receive IoT data, process and save it in the database in a generic way. The data is passed to the artificial intelligence, which produces specific responses. Then, the specific responses come back to the IoT. Organizing the code in auto-explained paths and sub-modules. Code related to a house in a path of home with its documents and code comments.
Best Practice CO9.1. Organize the folders logically, following a clear and organized structure. It is recommended to have a specific folder for each default setting, log, errors, and initialization script. In addition, it is essential to create separate folders and *subfolders* for different categories, such as documents, images, videos, and projects, for specific contexts. This approach makes it easier to locate and manage files. Implementing the code in such a reusable way and based on similarity ideas and code dependency.

Challenge CO9.2. Hierarchy and redundancy problems. During the development of the distributed system, practitioners faced challenges related to inheritance and code redundancy. For example, the IDs provided by third-party partners in various parts of the system caused code duplication. Third-party partner IDs are unique and universal, generated by the Universal Unique Identifier (UUID). In theory, a UUID is never repeated in the universe, but in practice, it may only be unique within the system.
Best Practice CO9.2. Configure the IDs that will be in each module. The IDs are intertwined in the modules. Because all the data is intertwined, a device is in a room, but that room is in a house. The house is connected to a user, and the user is connected to the house on both sides.

Challenge CO10. Working on software in progress. Receiving new members and training them to work in complex software, which is in progress and nearly to be implemented in a controlled environment.
Best Practice CO10. To study how architecture was developed. Continuous meetings and talking about small developments. Attribution of new tasks to the new members where these tasks can be developed externally.

Smart Module (SM): We present five challenges faced by developers of Smart Module and good practices for addressing them.

Challenge SM1. Data classification on the inputs and outputs of the connections between the Smart Module and the communications (COM) was a crucial point. Throughout the system, data is typed, but in the Smart Module part, it is formatted in JavaScript Object Notation (json). This entails certain specifications, such as transforming numbers into strings and *Booleans* into strings. This json behavior requires constantly checking the types of data received and sent. For example, suppose a user turns on a light bulb representing a new state about the previous state. In that case, the Smart Module will try to predict the next state of the light bulb and make a corresponding recommendation, which will be sent black to the core for processing.
Best Practice SM1. Save the typing used for transformation when returning the data.

Challenge SM2. Ensuring that the interaction and communication among AI models have a satisfactory result is complex.
Best Practice SM2. Brainstorming meetings to analyze the domain and develop ideas of what could be implemented. Then, test the best ideas and choose the one which has the best result.
Best Practice SM2. For the study, the practitioners used intelligence, which contains three abstractions. The intelligence that works for the environment, i.e., the house. The intelligence that works for the rooms and the intelligence that works for the devices. In this architecture, the intelligence that works for the environment connects with the intelligence that works for the devices, i.e., the environment predicts how the various devices will have their states modified, and through this modification, the smart module makes the prediction, it sends it, to these devices, in this case, the actuators. The actuators will predict the next state. While the room's intelligence works to predict the user's behavior, in other words, the intelligence predicts that there will be a change in some sensor in some device in the next few moments.

Challenge SM3. Real-time debugging of asynchronous functions.
Best Practice SM3. When the functions are asynchronous, it is hard to understand the results they deliver.

Challenge SM4.1. The challenge was the process of searching for the best algorithm and training to build the AI model for the development of the smart home that complies with the design of the third-party partners and the expectations of the planned smart home.
Best Practice SM4.1. Develop a simulation of an AI-based smart home using a model. In this Smart module, developers can use virtual devices/sensors to simulate a real environment.
Best Practice SM4.1. Base the modeling on the information provided by third-party partners on how we could use, retrieve, and send data to it.

Challenge SM4.2. Conduct research on machine learning algorithms and evaluate which have satisfactory accuracy.
Best Practice SM4.2. Compare the compression of predictions in Single and Multi-Label machine learning and determine which better fits the project.
Best Practice SM4.2. Compare machine learning models and identify the most satisfactory value based on the modeling. Our architecture uses machine learning based on neural networks in UMLP due to its satisfactory precision and ability to adapt to previously unseen behaviors.

Third-Party Partner (TPP): We present ten challenges and good practices reported by the Third-Party Partner developers.
Challenge TPP1.1. One of the project's most significant problems was that the third-party platform's message queue service needed to be updated, and the changes often interrupted communication with the project's devices. For example, the *bizcode "deviceonline"* was initially not available. The practitioners regularly monitored the changes to the API and then re-implemented the addition of the new functionalities in the sub-class so that the API updates did not directly interfere with the main code.
Best Practice TPP1.1. Create an intermediate class to make the necessary adaptations to the communication library with the API when there is an update and send it to the core after the update.

Challenge TPP1.2. The third-party partner API is constantly being updated. Some of the code we used to do would stop working after a while because the API had changed. Another point is that some of the functionalities were not even available. FOr example, when we create a room, add a device to the room, or add a user to the house, this information is not sent to the message queue service.
Best Practice TPP1.2. Unfortunately, the practitioners depend on the platform and its API updates. Nevertheless, check that the API documentation is being updated frequently. When it was written, it no longer worked at the time we were getting in the way because the API was being updated.

Challenge TPP2.1. Difficulty with the devices because we needed to have the devices directly from COMPANY A . For instance, each device has its specific

functionalities, and knowing them is important for the development and adaptation of the system. For instance, the function *switchled* can turn on or off a light bubble from a specific factory. However, another factory can use the *switchon* and *switchoff* functionalities for the same purpose.
Best Practice TPP2.1. Buy the devices from COMPANY A, and study each of the characteristics found in COMPANY A devices.

Challenge TPP2.2 Each device has different characteristics. For instance, the light bulb that turns on and off changes color and temperature. So, a light bulb has characteristics that are different from a plug.
Best Practice TPP2.2 The practitioners thoroughly examined each device and mapped the distinctive characteristics of the devices.

Challenge TPP2.3. The challenge was to choose the AI model so that accuracy would be satisfactory. Presently, many multi-label algorithm models exist, each distinguished by its unique set of characteristics.
Best Practice TPP2.3. The practitioners developed a script that systematically adjusts model parameters to find the model with the best accuracy and time. Afterward, the practitioners made a comparison between the machine learning algorithms. We selected the one with the highest satisfactory value depending on our modeling and achieved this accuracy in the initial test environments of over 98%. For this architecture, we used machine learning based on neural networks in Multilayer Perceptron (MLP) due to its satisfactory accuracy and ability to adapt to previously unseen behaviors.

Challenge TPP3. Difficulty in knowing energy consumption per device. Access to consumption information from devices to calculate energy consumption. For example, some devices, such as smart plugs, account for amperage above 45mA. Devices below 40mA do not account for consumption.
Best Practice TPP3. Know the real energy consumption of each device. To calculate the power consumption, we used a digital *multimeter* to measure the operational current of the device on standby. The value obtained was considered an offset, increasing the final sum.

Challenge TPP4.1. The practitioners used a third-party API platform. The platform was constantly evolving, so whenever we implemented something, it would break, and then we would turn to the documentation because we could not find it on the internet.
Best Practice TPP4.1. Searching for information in the API's documentation.

Challenge TPP4.2. The API documentation does not clearly show how the devices communicate with the third-party platform. Based on the tests, we assume that every time a smart device sends a message to the third-party platform or the third-party platform sends a message to the device, after receiving the message, it can connect to the device and only access the data after the fact, i.e., the light bulb needs to be switched on for the practitioners to receive information about the data.

Best Practice TPP4.2. The best practice to solve this delay was to access the information through third-party APIs called PULSAR. PULSAR is a framework (similar to MQTT) whose principal objective is to subscribe to a particular topic and listen when the devices communicate. So, the system should listen every time a device sends a message.

Challenge TPP4.3. Difficulty developing PULSAR concurrently. For example, PULSAR only works for one connection, although the documentation said it worked for several. PULSAR API did not allow connecting to multiple channels. PULSAR API was created fixedly to connect to only one channel, which is a standard channel, the default.

Best Practice TPP4.3. The solution was to create the project's API based on PULSAR API. The project's API allows connections to various channels and can communicate with PULSAR's AI.

5.2 Visualization (APP):

We present nine challenges faced by developers of the Visualization module and the good practices for addressing them.

Challenge APP1.1. Initially, during the phases of defining the technologies to be used, there was a communication gap between practitioners, project managers, and COMPANY A regarding the choice of technologies to be adopted. At the beginning of the project, there was no definition of whether the practitioner's team should use the same technologies as the COMPANY A team. The practitioners chose to use stable and widely used technologies in the industry. The practitioners spent three months perfecting some practitioner skills and implementing the base functionalities of the mobile application. Therefore, the practitioner's team had to translate the base functionalities considering new technologies, which demanded more time and effort for learning.

Best Practice APP1.1. Establish regular meetings to ensure continuous alignment on the use of technologies. To streamline the technology transfer process, the COMPANY A team proactively shared comprehensive documents outlining the technologies employed in mobile application development. Furthermore, the management team facilitated seamless communication between COMPANY A and the practitioners' teams, addressing queries related to the project's technological aspects.

Challenge APP1.2. There was a lack of communication between the teams during the integration of the modules concerning the modifications made, which harmed the development of the application. Before using the third-party APIs, the mobile application and core module communication was performed exclusively through a `json` for data requests. As both endings were in development, any changes in the core module would break the mobile application because it uses objects corresponding to json files.

Best Practice APP1.2. We chose to promote an effective conversation, and throughout the process, communication improved. The first solution consisted of improving the communication between the developers in the team. Every change in the core module that would affect the json files used in the mobile application was communicated through a specific channel. However, the modifications would take too much time and effort to co-evolve the mobile application interface and the core module. The developers implemented a specific server to intermediate the requests between the mobile application and the core module.

Challenge APP2.1. Integration and implementation of the project. First, the solution underwent adaptation to seamlessly integrate with the actual devices supplied by COMPANY A. Subsequently, the communication between the mobile application and the core module was managed through a third-party API instead of conventional requests. It led to an extensive rework on both fronts - the core module and the user interface.

Best Practice APP2.1. A good practice was to conduct a feasibility study with virtual devices and datasets. However, we could have simulated the requests more similarly to the third-party API to ease the translation from the feasibility study to the real-world application. Therefore, exchanging knowledge between the teams must allow developers to analyze the API documentation and other resources to make the simulations closer to reality.

Challenge APP2.2. Limitation of the technology used for developing the mobile application. The framework used to develop the mobile application has some limitations. For example, after requesting the core module, there was a constraint regarding the time and size of the message returned. The framework's documentation does not point out problems with long messages within requests and the error message is generic, making it hard to understand the problem.

Best Practice APP2.2. A solution is to work around the problems with the framework's limitations. In the specific case of the messages' size, the developers break the messages into smaller messages to exchange them between the core module and mobile application. As for other problems, developers found solutions in question-and-answer (Q&A) websites.

Challenge APP3.1. Problem with third-party partner documentation. The third-party API is constantly evolving. As a result, the documentation had some inconsistencies and missing information. For example, when creating the functionality to support many users in a home, it needed to be clarified whether they should be previously added as members or after accepting the invitation. Practitioners opt for the second solution, which might be better.

Best Practice APP3.1. The only solution is to check the documentation for updates constantly. Then, practitioners can adapt the code to match the best way of using the API methods.

Challenge APP3.2. Pairing problem, which caused the device to be restricted. The devices manufactured by COMPANY A have restrictions to connect with other unofficial mobile applications. It makes it hard to connect and test several devices in the context of the smart home.
Best Practice APP3.2. The problem was solved by COMPANY A team. COMPANY A team allowed the devices to connect to the practitioners' mobile applications for testing purposes.

5.3 DevOps:

We present six challenges and good practices reported by the DevOps developers.

Challenge DV1. `Kubernetes` documentation. The documentation provided needed to be more precise and provide a solution to the problems faced. The practitioners encountered limitations with online and cloud storage. To address this, we turned to NFS (Network File System), a distributed file system designed for local clusters. However, the `Kubernetes` documentation of NFS is incomplete, failing to provide detailed information on configuring these services.
Best Practice DV1. Practitioners explore through trial and error, looking for answers on QA platforms such as GitHub and online forums.

Challenge DV2. The first contact of practitioners with `Kubernetes` and `docker` may not be that difficult, but implementing a complete distributed system and integrating all modules with just a two-member group and limited expertise can indeed be quite challenging.
Best Practice DV2. Searching for forums and real examples helped us to understand how `Kubernetes` and `docker` work in a real development environment.

Challenge DV3. The complexity of communication between modules. Each module has unique computational processing, storage demands, network requirements, and dependencies. For each module, the practitioners have to sit down with the professional who developed the module and understand what each data entry and exit point looks like so that we can create the infrastructure for the modules.
Best Practice DV3. In collaboration with the practitioners of the other modules, the practitioners drew up a table describing each module's specific requirements. This process enabled the infrastructure to be adapted to each module's specific needs.

Challenge DV4.1. Communication between `nodes` and `pods`. During the integration test, the practitioners identified that the communication between `nodes` was unstable and sometimes provoked system interruption.
Best Practice DV4.1. The practitioners configured the digital certificate and set up communication between the `nodes`. Practitioners also requested that the

Information Technology department remove all the network blockages identified during the analysis.

6 Threats to Validity

In this section, we report the threats to validity according to Wohlin et al. [13].

Internal Validity. We interviewed 15 respondents, which is under the median for qualitative studies in software engineering (30 respondents). Besides, we kept the interview duration short to mitigate fatigue.

Construct Validity. Although some questions are general regarding their focus on software development challenges and best practices, we guide practitioners to answer in the specific context of smart homes. In addition, we asked follow-up questions to clarify the participants' responses, ensuring that we accurately captured their experiences and perspectives.

External Validity. As we interviewed practitioners from the same company, our results may not adequately represent the practices adopted in the wider software engineering industry. There is an agreement among the practitioners' responses, which reveal promising insights into the Brazilian scenario.

Conclusion Validity. For the data analysis, we transcribed the recorded interviews and analyzed the transcriptions line-by-line. The data analysis was an exhaustive process, which depended on the researchers' interpretation of the data. To minimize the chance that preconceived notions about the topic influenced the researcher, one researcher analyzed the data, and another checked the classifications. All the disagreements were discussed among all authors.

7 Conclusion

This study investigated the challenges and practices for developing smart homes through interviews with 15 practitioners from a Brazilian company. The practitioners' different backgrounds allowed us to identify challenges and good practices for different modules of a smart home application.

The challenges ranged from optimizing system performance to managing latency, data modeling complexities, and communication problems between development teams. For example, Core was one of the modules where we found the most significant challenges, such as optimization, latency, and data modeling. Best practices included considerations of distributed systems, real-time processing, and effective communication within the development team. Similar challenges and solutions were outlined for other modules, each one adapting to the requirements and complexities of the components.

As the project evolved, the team has shown adaptability and resilience, emphasizing the importance of continuous learning and collaboration among developers while developing smart homes. The challenges have provided insights into practices that can serve as lessons for future projects, contributing to the growth and maturation of the team and its capabilities. In future work, we will explore partnerships with third parties to integrate innovative devices and additional functionalities into the existing system. In addition, we will validate the good practices proposed in this paper.

Acknowledgments. This work was funded by Positivo Tecnologia and Fundação de Amparo à Pesquisa e Extensão (FAPEX) under grant number 391/2022, Fundação de Amparo a Pesquisa do Estado da Bahia (FAPESB) grants BOL0599/2019 and PIE0002/2022. This work is partially supported by INES (www.ines.org.br), Conselho Nacional de Desenvolvimento Científico e Tecnológico (CNPq) grants 312195/2021-4, 465614/2014-0, and 313623/2023-6, CAPES grant 88887.136410/2017-00, and FACEPE grants APQ-0399-1.03/17 and PRONEX APQ/03881.03/14.

References

1. Interview Script - Challenges and best pratices in smart home developing. 0 **0**(1), 1 (2024). https://figshare.com/articles/dataset/Interview_Script_-_Challenges_and_best_pratices_in_smart_home_developing_/26206793
2. Carra, C., Tabia, K.: Smart home for seniors: Opportunities and challenges for AI. In: ICAART 2020, 12th International Conference on Agents and Artificial Intelligence (2020)
3. El-Azab, R.: Smart homes: Potentials and challenges. Clean Energy **5** (2021)
4. Fernandes, E., Rahmati, A., Jung, J., Prakash, A.: Security implications of permission models in smart-home application frameworks. IEEE Secur. Priv. **15**(2), 24–30 (2017)
5. Gottfried, I.B.B., Aghajan, H.: The praxis of cognitive assistance in smart homes. In: Behaviour Monitoring and Interpretation-BMI: Smart Environments **3**, 183 (2009)
6. Makhadmeh, S.N., Khader, A.T., Al-Betar, M.A., Naim, S., Abasi, A.K., Alyasseri, Z.A.A.: Optimization methods for power scheduling problems in smart home: Survey. Renew. Sustain. Energy Rev. **115** (2019)
7. Ramos, C., Augusto, J.C., Shapiro, D.: Ambient intelligence-the next step for artificial intelligence. IEEE Intell. Syst. **23**(2), 15–18 (2008)
8. Robles, R.J., Kim, T.H.: Applications, systems and methods in smart home technology: A. Int. J. Adv. Sci. Technol. **15**, 37–48 (2010)
9. Sikder, A.K., Babun, L., Aksu, H., Uluagac, A.S.: Aegis: a context-aware security framework for smart home systems. In: Proceedings of the 35th Annual Computer Security Applications Conference, pp. 28–41. ACSAC '19, Association for Computing Machinery, New York, NY, USA (2019)
10. Singh, B., Khan, M.Z., Senthil, J.: Applications and challenges in IoT based smart homes. Int. J. Mech. Eng. **6** (2021)
11. Stol, K.J., Ralph, P., Fitzgerald, B.: Grounded theory in software engineering research: a critical review and guidelines. In: Proceedings of the 38th International conference on software engineering, pp. 120–131 (2016)
12. Thakur, P., Goel, S., Puthooran, E.: Edge AI enabled IoT framework for secure smart home infrastructure. Procedia Comput. Sci. **235**, 3369–3378 (2024)

13. Wohlin, C., Runeson, P., Hst, M., Ohlsson, M.C., Regnell, B., Wessln, A.: Experimentation in Software Engineering. Springer Publishing Company (2012). https://doi.org/10.1007/978-3-642-29044-2
14. Zhu, J., Wang, D., Zhao, Y.: Design of smart home environment based on wireless sensor system and artificial speech recognition. Measurement: Sensors **33**, 101090 (2024)

Party Without a Cake? Onto an Inter-modal HitchHike Logistics Platform for Passengers and Products Transportation

Mohammed Fahad Ali[✉], Dominique Briechle[✉], Marit Briechle-Mathiszig[✉], Tobias Geger[✉], and Andreas Rausch[✉]

Institute for Software and Systems Engineering, Clausthal University of Technology, 38678 Clausthal-Zellerfeld, Germany
{mohammed.fahad.ali,dominique.fabio.briechle,marit.elke.anke.mathiszig,
thomas.tobias.marcello.geger,andreas.rausch}@tu-clausthal.de

Abstract. Transportation is essential for facilitating the efficient movement of people and products, yet confronted by problems including congestion, high operating expenses, and negative impact on the environment. Inter-modal transportation provides a way to mitigate these issues through reduced environmental impact and increased logistical efficiency. However, major obstacles are preventing its widespread implementation, including the integration of multiple systems, the requirement of supporting infrastructure, and the substantial cost of integration. Thus, fostering an open and distributed ecosystem for different service providers supports and optimizes inter-modal transportation. Nevertheless, offering openness and distribution compromises system reliability due to the high level of complexity and variability of service integration among different modes of transportation and independent operators. This paper proposes the architecture of the HitchHike logistics platform that supports inter-modal multi-hop transportation segment for products and passengers by hitchhiking through the local hubs and aims to balance system openness with reliability. In addition, the utilization of artificial intelligence (AI)-based dynamic real-time route planning facilitates efficient routes and aims to optimize the efficiency of individual modes of transportation while maintaining flexibility. According to the customer's anticipated risk, the system offers flexible routes which correspond to the expected risk. Therefore, through an integrated ecosystem of logistic-based service providers, the proposed architecture addresses challenges including navigating various transportation networks, choosing efficient and economical routes, and allowing key stakeholders to collaborate and work proactively.

M. F. Ali, D. Briechle, M. Briechle-Mathiszig, T. Geger and A. Rausch—These authors contributed equally to this work.

© The Author(s), under exclusive license to Springer Nature Switzerland AG 2024
A. Ampatzoglou et al. (Eds.): ECSA 2024, LNCS 14937, pp. 100–114, 2024.
https://doi.org/10.1007/978-3-031-71246-3_11

Keywords: Inter-modal Transportation · Dynamic Re-routing · Platform Architecture · Logistics Service Management · Sustainable Logistics

1 Introduction and Motivation

In today's interconnected world, the transportation network is a necessary component of our contemporary society. Transportation is essential for connecting people, products, and services that drive economic growth and enhance the quality of life. However, the progressive growth in commercial logistics-based transportation has impacted land utilization, leading to numerous problems including a steady increase in traffic congestion, longer commutes, parking challenges, and increased CO_2 emissions. In addition, these problems are being exacerbated by the population growth and increasing migration to the metropolitan areas [9].

According to the European Union (EU) Commission, there will be an approximate projection of a fifty percent increase in traffic congestion by the year 2050, with road transportation continuing to be crucial to this freight movement [7]. In addition, another estimation indicates that traffic congestion costs the economy of the EU about two percent of its gross domestic product [9]. Consequently, the EU is already implementing several sustainable mobility initiatives to minimize the negative environmental effects of energy consumption and maximize the effectiveness of the current transportation system [7].

One strategy adopted by the EU to disentangle mobility from its drawbacks is the encouragement of inter-modal transportation [7]. Inter-modal transportation uses a chain of networks involving at least two different modes and services of transportation from the origin to the destination [5]. In addition, synchronizing various modes of transportation is essential in achieving short and long-term decarbonization targets [1].

However, the EU Commission has noted certain barriers and costs of friction that impede inter-modal transportation. For instance, the lack of a cohesive network of modes and their interconnections, and the incompatibility of various modes of transportation [7]. In addition, along with a wide range of decision-makers and stakeholders, different operations and planning activities are all involved in an inter-modal transportation network [5]. Therefore, logistic providers encounter unexpected challenges when utilizing inter-modal transportation networks [1].

Additionally, the unpredictability and variances in transportation networks can stem from multiple sources, such as planned service interruptions, weather-related risks, or congestion on interconnecting routes. Therefore, the reliability of these networks is compromised as all the decisions in transportation planning are based on the theoretical aspects of future operations. As a result, not all unexpected circumstances occurring in real-world scenarios can be managed by probability-based planning strategies [14]. Moreover, the optimization of inter-modal transportation networks requires proactive engagement and effective collaboration among different service providers. However, most studies neglect the possibility that different service providers may have conflicting objectives when it comes to inter-modal transportation [2]. Consequently, according to Giuffrida et al. [8], to achieve an intelligent and sustainable inter-modal transportation system, it is imperative to concentrate on shifting the system architecture from a closed to an open and integrated system for different service providers. The following research questions constituted the main emphasis of this paper:

1. How might an architecture be designed that balances openness to integrate different service providers with reliability and promote inter-modal transportation for both products and passengers?
2. How can an end-to-end transportation network, in a decentralized manner, dynamically adjust to re-planning and accommodate routing changes to maximize efficiency and reliability?

The paper is organized as follows. Section 2 provides a succinct overview of the state-of-the-art research. Section 3 outlines the scenario focusing on current circumstances and impending issues. The same scenario utilizing the proposed solution is described in Sect. 4. Further, Sect. 5 covers the proposed architecture of the logistics platform. Finally, Sect. 6 summarizes the experience with the HitchHike system followed by Sect. 7 with the conclusion.

2 State of the Art

Worldwide research and attention are increasingly focused on inter-modal transportation as a more environment-friendly strategy to mitigate climate change. De Miranda Pinto et al. [13] conducted a case study to evaluate the viability of an inter-modal network that combines rail and road transportation. The outcomes of the case study demonstrate that the inter-modal network utilizing road and rail transportation significantly outperforms uni-modal road transportation concerning efficacy and effects on the environment. Specifically, it can reduce emissions by about seventy-seven percent, enhance energy efficiency by approximately forty-three percent, and decrease operating expenses by as much as eighty percent. Nevertheless, despite these advantages, integrating several systems and stakeholders is the major challenge confronted by inter-modal transportation [8]. In addition, the findings of the European project on inter-modal quality indicate that except for a few major flow routes, inter-modal transport in Europe generally performs inefficiently overall [4]. Also, the inter-modal transport policy by the EU has not achieved substantial progress [3].

In the context of inter-modal transportation, route planning is frequently concerned with mitigating disruptions and ensuring the successful execution of transport tasks. Thus, a model-based strategy for freight transportation in the hinterland is presented by Qu et al. [14]. The constructed model controls the re-scheduling of services and re-routing of freight movements to re-plan the given freight transportation. However, the recommended transportation plan is unsustainable and requires too many modifications from the existing plan [1]. In addition, the model neglects to account for the impact of delays and congestion brought on by overloaded terminals [14].

Consequently, Akyüz et al. [1] suggest a re-planning-based decision tool for transportation to handle potential disruption scenarios and concentrate on the re-planning of the shipment flow in an inter-modal network. However, the approach is restricted to the number of trans-shipments and delivery time-frames specified by the customer. According to the authors, further research can address other disruptions including changes in capacity, demand, and delivery time constraints. In addition, another direction of research can include products with different priorities, such as products requiring cold chain transportation, or costly and delicate goods.

Moreover, Darayi et al. [6] emphasize adaptive capacity planning strategies to effectively manage the supply-demand network in case of an unexpected event, to minimize the economic impact by using strategic re-routing mechanisms. A case study on freight transportation planning is used to demonstrate the presented approach, which tries to balance multiple objectives while improving risk management strategies. However, it takes extensive research to determine the most efficient solutions while balancing multiple objectives as it is complex in terms of modeling [2]. In addition, the presented approach neglects the conflicting nature of multiple stakeholders interacting in an inter-modal network or various negative external events as an objective to be balanced [2].

According to a survey by Archetti et al. [2] to investigate different challenges and optimize freight transportation, the approaches that consider interconnected and synchronized systems of various modalities still need to be thoroughly investigated. In addition, inter-modal network integration can be broadened by focusing on issues concerning hub management for inter-modal terminals and scheduling services for equipment handling. The authors emphasize that the objective to strive towards in the future is the dynamic optimization of shipment flows by incorporating various modalities and focusing on stakeholders to coordinate and work proactively.

Giuffrida et al. [8] presented some of the outcomes of the "Intelligent Transport Systems (ITS) Italy project 2020" including the development of a prototype solution for improving the monitoring of freight transport throughout the inter-modal transportation network. In addition to providing real-time freight visibility, tracking, and data collection, the prototype can communicate with sensors and devices installed onboard. The authors emphasize that it is essential to focus on shifting the information and communication systems from closed to an open and integrated system architecture to attain a sustainable inter-modal transportation system.

Most notably, the concept of "Hitchhiking" has already led to a number of publications, describing different states and sections of the anticipated system. Initially, the idea of "Hitchhiking" smart devices in this context was presented by Lawrenz and Leiding [12]. The paper centers around the question of a systematic approach to enable smart devices to transmit information and therefore carrying for their own disposal at the final End-of-Life (EOL). The idea of the system was at that time, that the smart devices were conducting this process on their own, including the conduction of payment and the booking of the pickup task at a specific logistic provider. The hereby created smart contract is then stored on a Blockchain, allowing tracing of the process.

In addition, the Hitchhike concept focusing on data policy, security, and encryption of cyber-physical systems is researched by Werner et al. [15]. The paper describes the requirements in the context of medical transportation with the help of a smart Internet of Things (IoT)-based system with an emphasis on data storage. The requirements for such a system are stated as well by [15] laying a high emphasis on system resilience, query-ability, and computing power. It proposes a system, where data is stored on distributed databases, which has certain benefits of the reduction of participant's trust.

Finally, the context of emergent platform systems for smart mobility applications was investigated in this context by Wilken et al. [16]. Especially with the emphasis on the sustainability of software components, the paper leverages the idea of re-usability of emergent sequences of software artifacts. Based on the initial requirement of the user, a planning problem is created. The specified domain will consecutively be used in the solving process to set up a sequence of software components, matching the initially described requirements in the problem. In this way, the conception of smart mobility systems can be done based on user requests.

3 Scenario

In order to describe the general obstacles of currently in-use logistic systems, we defined a short and comprehensive scenario, which helps the reader to understand the barriers of those systems from a user perspective. The scenario therefore tries to lay emphasis on the problems of the current state of individual transportation service and will be replayed with the system-wise solution the HitchHike-System offers in the Discussion section of the paper. For our scenario, shown in Fig. 1, we imagine a kind of everyday situation. Our central protagonist in the scenario is Jessica. Jessica is living in a small village (S3) and is planning to celebrate today her 40th birthday in the town hall (E2) of a nearby city. This birthday should be one of a kind and Jessica planned a huge party with her friends and family.

However, so far this was not a happy day at all for Jessica. This morning her car broke down and she had to call the bakery (S2) and the party outfitter (S4) and tell them, that she was unable to pick up the goods she ordered. The companies have therefore planned to deliver the goods by themselves with additional cost for Jessica. On top of that, Jessica now needs transportation for herself and is uncertain, if she wants to walk, which is without additional cost, or if she wants to take a cab, which is more convenient but comes with extra cost. She finally decides to take a cab to her party. Her mood is however already at a low and the party is spoiled. Simultaneously, Sarah (S1) is driving from her house to meet up with her friend Layla (E1). If only there would be a solution which could have solved the issue and combined the transportation ways...

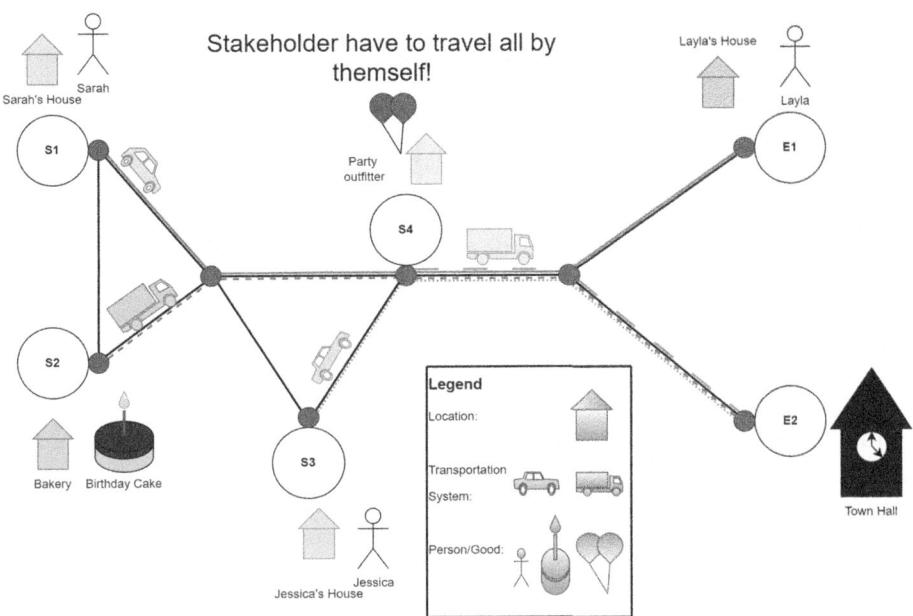

Fig. 1. Setting of the Scenario with individual transportation to one common location

This small scenario shows, how transportation, especially individual transportation, in rural areas can be quite difficult, especially with a lack of personal vehicles. Stakeholders often need to rely on individual transportation or costly services. Often it would therefore be possible to consolidate transportation tasks and by that not only save cost for all stakeholders but also reduce the ecological impact on logistics for personal and goods.

4 Solution

The presented HitchHike system could help our initial protagonist Jessica greatly. We have therefore updated our scenario in Fig. 1 with the implemented HitchHike system which our personas now can use to solve their initial dilemma.

In the alternate scenario, Jessica (S3) is a long-term user of the system and has registered even her house as a location in the system's network. She therefore decides to order her cake (S2) and the decoration (S4) for the party online at the respective shops of the suppliers. Both of them are luckily as well users of the Hitchhike-System and therefore create a specific pick-up task for the delivery of their goods. Simultaneously, Jessica decides to use the HitchHike System herself and creates a pick-up task for herself to get to the party. Sarah (S1) is registered as a driver in the HitchHike System and is driving today from her house to her friend's house Layla (E1). She is commuting every day between the two locations so she accepts the request. The bakery, Jessica, and the party outfitter are now receiving the notification, that a driver has been matched to their transportation request. Layla is departing at 4:00 p.m. and is driving to the bakery to pick up the cake and transits afterward to Jessica's home, where she is going to pick up Jessica as well. Afterward, they travel to the next location of the routing service prognoses, which is the hub located at the party outfitter. They are picking up the goods for the party and are then heading to the next hub, which is near Sarah's friend's house Layla. But misfortune for Jessica has not stopped yet. Layla and Jessica are caught in traffic for over an hour, which is now messing up Jessica's plan for the evening. The ride is taking way longer than expected and Jessica is afraid, that she will miss her party. Layla drops Jessica off at the hub and continues her journey to her friend's house Layla. The original available adjunct ride at 4:30 p.m. is now already gone. She therefore gets the notification, that Paul will pick her up with his electric rickshaw as displayed in Fig. 2. Since Paul is the only one left commuting today between the hub and the town hall (E2), he has received all three of the transportation requests from the bakery, the party outfitter, and Jessica. However, since Layla was already very late, Paul's delivery conduction would be at high risk of being too late to the party.

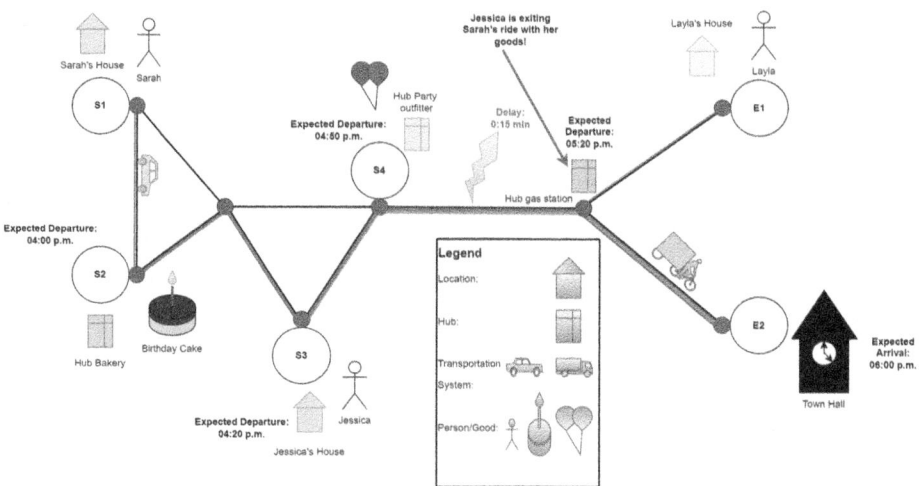

Fig. 2. Setting of the Scenario with the marked distance between the different locations for a high logistic risk.

Jessica now has the option to call an "emergency" mobility provider as described in Fig. 3. This provider would be more costly but would minimize the risk of being too late drastically. She finally selects the more costly option and the ride is therefore rebooked. The bakery and party outfitter have as well the opportunity to use now faster transportation without additional cost, which they are accepting as well. Jessica, her cake, and the decorations are arriving at 5:35 p.m. without any further delay and can now start fully relaxed into a hopefully enjoyable evening.

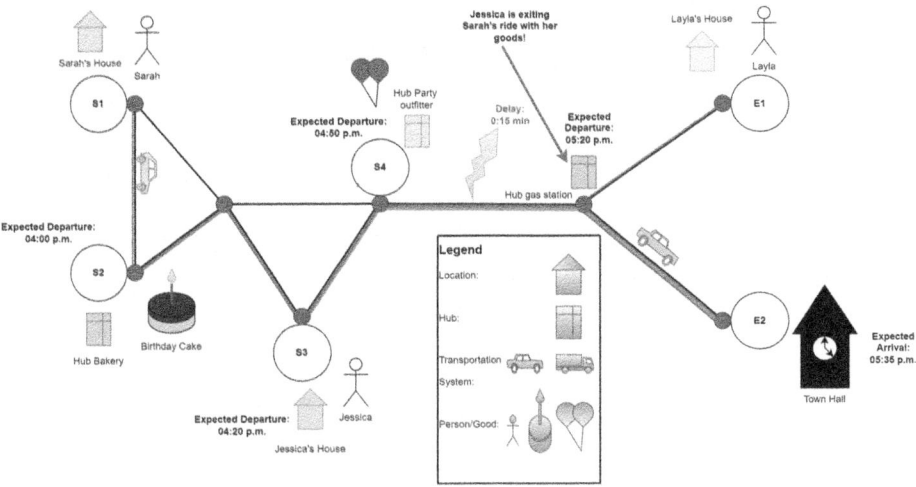

Fig. 3. Setting of the Scenario with the marked distance between the different locations for a low logistic risk

The solution scenario shows, how intelligent, AI-based route planning can adapt to specific circumstances. Not only does it ensure, that all the stakeholders receive suitable transportation for a given good, but in addition it enables an economical and ecological logistic by aligning different mobility necessities. In our scenario, the adaptability of the system is shown in terms of route optimization. Each planning cycle finds the best route from those suggested by AI-based dynamic real-time route planning since it considers the user's acceptable level of risk and the initial delivery requirements. If this is not possible within the delivery requirements, an "emergency" transportation can be dispatched.

5 Platform Architecture

The degree of uncertainty and complexity of service integration among various transportation modes compromises system reliability, which limits offering an openness to integrate different service providers. The key premise of the open service-based HitchHike logistics platform is to support inter-modal transportation for products and passengers and allow key stakeholders to collaborate and work proactively while balancing the system reliability with the openness to integrate different service providers, as illustrated by the platform architecture in Fig. 4.

The user interacts with the platform for private or business-related logistics by providing the requirements and the preferred degree of risk to the logistics task management service. This component manages the logistical requirements for products and passenger transportation while focusing on balancing risk and cost. The user can designate cost or time as the optimization criterion for the

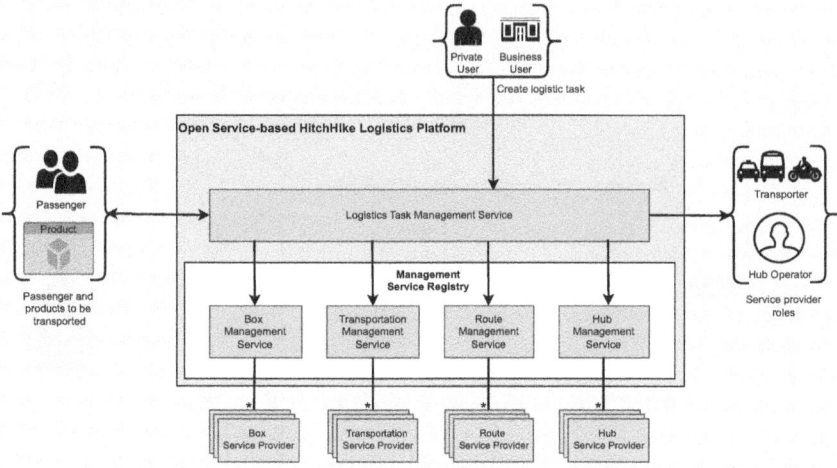

Fig. 4. Open service-based HitchHike logistics platform architecture

logistic task. The logistics task management service processes the user requirements and interacts with the passenger or product that needs to be transported along with the corresponding service provider for the assigned role in each transportation segment. In addition, the services that are managed and usable through the platform by different service providers are included in the management service registry. The management service register answers the question concerning who are the different service providers that can plug into the HitchHike logistics platform and what services they can offer in an integrated manner. Moreover, the management service registry also manages the way of interaction with the different service providers. In addition to having different kinds of service providers connected to the platform, multiple service providers can fulfill each management service that is offered. This is illustrated by the star symbol notation in Fig. 4, which represents the cardinalities according to the Unified Modeling Language notation.

Most notably, the reliability of the system is balanced with the openness offered by the platform to integrate multiple service providers of different kinds. For instance, the box provider provides the "HitchHike box", which are reusable logistics boxes with IoT components utilized for product transportation. Thus, multiple box providers can plug into the platform and provide the boxes resulting in product readiness and scheduled pickups. The same applies to the transportation service providers which offer multiple mobility options for inter-modal transportation of products and passengers from start to end hub. In addition, the platform also allows connecting with multiple route providers and allows the integration of different optimization methods for routing, which optimizes routes for efficiency and travel times, whether for shorter passenger journeys or longer product routes. Finally, the hub service provider manages hubs where people can transit between various transportation modes and products can be temporarily stored, or transferred between modes of transportation.

When it comes to handling the logistics task, the logistics task management service creates an agent for each task that the user initiates. The created agent requests the potential route plans for each transportation segment after completing the previous segment. The connected route providers utilize AI-based dynamic real-time route planning to estimate the potential route for each transportation segment by analyzing the existing traffic forecast training data of the fleet vehicles [10]. Next, the potential route plans are assessed and ranked by the agent created for the assigned task. Consequently, the route plan that best fits the optimization criterion and user's desired level of risk is selected. Subsequently, the agent makes a request to the connected transportation and hub service providers and books the transporter with the start and end hub for the selected route. Finally, the agent created for the assigned task coordinates the route plan and transmits vital information between the passengers, transporters, and hub operators concerning the planned transportation segment. This process is repeated until the destination is reached. In the event of an unforeseen circumstance, the agent can immediately perform re-planning as all the potential routes are already evaluated and ranked based on a set of optimization criteria including cost and time as well as the acceptable level of risk as predefined by the user.

In the case of product transportation, the Hitchhike box has integrated functionality that uses IoT-based components to continuously monitor the product. As a result, the assigned agent is connected to the box and receives real-time information based on the environmental conditions at every stage of the intermodal journey. This is significantly important for products like food and sensitive medical supplies because of several factors, including inadequate storage conditions, exposure to air or sunlight, and the degree to which these products are susceptible to environmental changes while in transit across different modes of transportation.

The overall cost of fulfilling the logistical requirements of the user is inversely correlated with the accepted degree of risk. The low-risk options are comparatively more expensive but ensure higher dependability. Conversely, the high-risk options are relatively cheaper but offer less assurance about timeliness and entail a chance of possible congestion. In the case of low-risk scenarios, the user expects minimal disruption and thus the route circumstances are continuously monitored and re-planning is initiated, if the planned route becomes delayed or obstructed above a predetermined level. This could entail adjusting to alternative route plans or changing modes of transportation based on real-time data and emergent conditions, with the aim of guaranteeing system reliability for onboard users to reach their destination.

Furthermore, the shared responsibilities among the plugged service providers ensure that each service provider maintains and enhances the overall effectiveness and reliability of the platform. For instance, transportation service providers are particularly accountable for efficiently managing transporters, as they are pivotal to the management of direct transportation segments. Consequently, transportation service providers have the freedom to accept or decline ride requests based on real-time capacity and demand estimates to avoid situations like overbooking. This sharing of responsibility is essential in reducing inefficiencies and enhancing customer experience since it prevents the discomfort and difficulty that come with circumstances like overbooking of rides. In the same way, hub providers are responsible for overseeing the logistics infrastructure at the designated hubs and are accountable for enabling smooth transitions between different modes of transportation in the event of delays and congestion. Therefore, every service provider connected to the logistics platform has distinct responsibilities that are essential to the efficient functioning of the transportation network. Lastly, in the event that the requested service provider declines the request, the system will either choose the best possible alternative or immediately start re-planning to explore for other possibilities.

6 Experience with the HitchHikeBox System

Selected parts and variants of the HitchHikeBox System were already demonstrated at several places. For example, a system variant for the transportation of products and passengers between the cities Clausthal and Goslar in Germany was demonstrated. Hereby the sender generates a delivery task in our web user interface and deposits the transportation box, which is called the "HitchHikeBox", in a hub in Clausthal. After that, a driver took the HitchHikeBox and drove to an outdoor hub in Goslar namely Goslar-Outside. Next, the HitchHikeBox is fetched up by another driver, who further delivers the box from the hub Goslar-Outside to the hub Goslar-Inside. The whole process including several steps is further described in more detail:

At first, the sender generates a delivery task for the product in our web user interface, using the preferred start and end hub and defining the delivery requirements as well as stating the accepted price. Afterward, the sender puts the product in the HitchHikeBox and deposits the box in a hub case that can be assessed by using a QR Code that was sent via a mobile app to open the hub case. The utilized hubs are constructed similarly to the package stations of common package delivery services. After confirming the box drop-off, the first route segment was planned. A driver who drives from Clausthal to Goslar for an appointment received a notification that he could pick up and deliver the HitchHikeBox. He accepted the transportation task and therefore received payment for the conducted transportation route. Additionally, he received a QR Code to open the hub case, got the HitchHikebox, and delivered the box to the hub Goslar-Outside.

At the hub Goslar-Outside, the driver opened the hub case, dropped off the HitchHikeBox, and confirmed the drop-off. The next route is planned and a cyclist who rides his bike from hub Goslar-Outside to hub Goslar-Inside receives a notification via the mobile app. He accepted the transportation task, opened the hub case, and received the HitchHikeBox. Afterward, the cyclist delivered the Box from the hub Goslar-Outside to the hub Goslar-Inside. The hub case was also opened here, the HitchHikebox was placed inside and the drop-off was confirmed. The receiver got a notification concerning the pick-up of the product. Finally, the receiver opened the hub case and received the HitchHikeBox with the product.

This partial implementation of the HitchHikeBox System was demonstrated and filmed. The produced video is accessible through the URL provided in reference [11] and Fig. 5 shows certain steps of the scenario, including the storing of a product in the HitchHike Box, the storage of the Box in a Hub system, the inter-modal aspect of the system, as well as the final usage of the product by the recipient.

Fig. 5. Expressions of the HitchHike logistics Use-Case [11]

The HitchHike logistics platform proposes a systematic concept for an inter-modal logistics platform system, which enables different stakeholders to link specific services into the interfaces of the platform. This conception offers a new approach to inter-modal transportation in general by offering the alignment of requests for passenger as well as product transportation. Systematic solutions like the proposed conception can therefore help solve logistic issues in cities as well as rural communities by offering a far wider variety of possible transportation solutions in a consolidated system. Current logistic distributors, especially in the domain of public transportation can link into the system just as well as private citizens, who are commuting between different locations, therefore offering the chance to reduce further traffic by aligning such commuting tasks. Besides these alignment factors, the system allows the harmonization of passenger and product transportation, which offers a high degree of flexibility for the stakeholders, who are posting a package and are normally relying on a few companies that are offering limited flexibility in terms of time and location to send a package. The linkage of different service providers further minimizes the risk of a "blackout" of rides, meaning that there is no potential transportation available. With the integration of a manifold of different stakeholders, the risk mitigation practice becomes better with the number of potential providers. Simultaneously, the economic incentive for mobility providers is increasing as well with the number of end-users conducting their tasks via the HitchHike system. The proposed system emphasizes in this are a high degree of responsibility to their senders, not only in terms of trust but also in terms of self-organization. This is however also one of the potential weak spots of such a system. Since the transportation providers have the ability to self-organize the booked routes directed to a specific provider, the end-user who triggered the ride has to rely on the competence and ability of the provider to manage the booked rides on their end. This potential drawback however has as well its upsides: With shared responsibility, the potential contribution for mobility providers is more interesting, because it allows them to maintain a certain degree of self-management of their fleet. This especially comes into place for special deliveries which are difficult to handle based on their attributes.

In terms of future research directions, the management of the ride alignment via the system rather than the mobility provider could offer a viable foundation for extending the system. The open platform architecture enables the extension of the system at different points, leveraging the further development of additional services which are supporting the optimization of mobility services. Moreover, dynamic scenario handling could be further researched which enables the system to react to different situations in a more granular way, therefore enabling a higher grade of autonomy and increasing user satisfaction.

7 Conclusion

The proposed logistic system offers a drastic enhancement of transportation management while simultaneously reducing the cost for all of its stakeholders by aligning transportation tasks. The concept is set up as a modular system to enable the docking of different service providers as depicted in the system architecture in Fig. 4 and therefore mitigates the risk attached to a specific provider by offering alternatives, therefore answering the question of reliable and open system concepts for inter-modal logistic systems.

Moreover, the proposed architecture concept enables the adaptability of the sub-modules in terms of their adjustment to the dynamical re-planning of the reroutes. This is encompassed by a governing entity, namely the agent of the delivery task, responsible for selecting the proper service provider response. The proposed architecture could therefore address both our initial research questions and open up the topic of adaptable logistic systems for further research. The scenarios showed the application of such a decentralized system approach and the handling of the trade of between high risk/ low cost as well as low risk/ high cost. The currently implemented features of the systems were further evaluated in a demonstrator use-case and provided the first functionalities in terms of inter-modal logistics management with different transportation systems. The presented concept is therefore a valuable contribution in terms of highly flexible and risk-mitigating logistic systems, which offer a high level of scalability.

Acknowledgement. This work was funded by the German Federal Ministry for EconomicAffairs and Climate Action (BMWK) (Research Grant: 01ME-21002, Project: HitchHikeBox). We also thank the Developers of the Team J. Boll, H. Hemmerling and B. Raffie for their work regarding the implementation of the concept.

References

1. Akyüz, M.H., Dekker, R., Azadeh, S.S.: Partial and complete replanning of an intermodal logistic system under disruptions. Transp. Res. Part E Logistics Transp. Rev. **169**, 102968 (2023)
2. Archetti, C., Peirano, L., Speranza, M.G.: Optimization in multimodal freight transportation problems: a survey. Eur. J. Oper. Res. **299**(1), 1–20 (2022)
3. European Court of Auditors: Rail freight transport in the EU: still not on the right track (2016)
4. Cardebring, P., Fiedler, R., Reynauld, C., Weaver, P.: Summary of the IQ project. Analysing intermodal quality; a key step towards enhancing intermodal performance and market share in Europe (2000)
5. Crainic, T.G., Perboli, G., Rosano, M.: Simulation of intermodal freight transportation systems: a taxonomy. Eur. J. Oper. Res. **270**(2), 401–418 (2018)
6. Darayi, M., Barker, K., Nicholson, C.D.: A multi-industry economic impact perspective on adaptive capacity planning in a freight transportation network. Int. J. Prod. Econ. **208**, 356–368 (2019)
7. Eftestøl, E.J., Bask, A., Rajahonka, M.: Intermodal transport research: a law and logistics literature review with EU focus. Eur. Transp. Law **49**(6), 609–674 (2014)
8. Giuffrida, M., Perotti, S., Tumino, A., Villois, V.: Developing a prototype platform to manage intelligent communication systems in intermodal transport. Transp. Res. Procedia **55**, 1320–1327 (2021)
9. Guerrero-Ibanez, J.A., Zeadally, S., Contreras-Castillo, J.: Integration challenges of intelligent transportation systems with connected vehicle, cloud computing, and internet of things technologies. IEEE Wirel. Commun. **22**(6), 122–128 (2015)
10. Project Website "HitchHikeBox", University of Mannheim. https://www.uni-mannheim.de/ines/projekte/projektuebersicht/hitchhikebox/. Accsessed 09 Jul 2024
11. Project Video "HitchHikeBox-System", funded by the federal ministry for economic affairs and climate action. https://www.youtube.com/watch?v=EAzjUiDtPgc. Accsessed 16 Jun 2024

12. Lawrenz, S., Leiding, B.: The hitchhiker's guide to the end-of-life for smart devices. In: International Symposium on Software Engineering for Adaptive and Self-Managing Systems (SEAMS), pp. 196–202 (2021)
13. de Miranda Pinto, J.T., Mistage, O., Bilotta, P., Helmers, E.: Road-rail intermodal freight transport as a strategy for climate change mitigation. Environ. Dev. **25**, 100–110 (2018)
14. Qu, W., Rezaei, J., Maknoon, Y., Tavasszy, L.: Hinterland freight transportation replanning model under the framework of synchromodality. Transp. Res. Part E Logistics Transp. Rev. **131**, 308–328 (2019)
15. Werner, R., Briechle, D., Mathiszig, M.: Hitchhikebox: a decentral, verifiable, and privacy-protecting automated logistic transport concept for pharmaceuticals. In: International Conference on Computer Technology Applications, vol. 10 (2024)
16. Wilken, N., et al.: Emergent software service platform and its application in a smart mobility setting. In: ADAPTIVE 2023 : The Fifteenth International Conference on Adaptive and Self-Adaptive Systems and Applications, vol. 15, pp. 11–14 (2023)

Development of Blockchain Network for Quality and Product Safety Control Information System

Aneta Poniszewska-Marańda(✉), Michał Pawlak, Maciej Kopa, and Mateusz Owczarek

Institute of Information Technology, Lodz University of Technology, Lodz, Poland
{aneta.poniszewska-maranda,michal.pawlak}@p.lodz.pl, 230460@edu.p.lodz.pl

Abstract. SALUS is an IT system that collects information about everyday products and their users' opinions in order to control their quality and detect potential double quality of products. This task requires a stable and reliable data "store" that is trusted by users and subsystems using it. Building such trust requires transparency of operation and resistance to manipulation of the entire system, including interactions between individual modules. The paper presents the SALUS.BC component that is the central module of the SALUS product quality and safety control system. The main element of the module is a distributed and decentralized blockchain network based on the Hyperledger Fabric distribution. It is an open-source implementation of blockchain technology provided by Linux Foundation. It offers an enterprise version of blockchain technology that relies on public and verified participants who reach consensus using the Raft algorithm. Hyperledger Fabric also offers support for smart contracts created using general-purpose programming languages and an advanced logical security model.

Keywords: Software architecture · service oriented computing · Blockchain technology · Hyperledger Fabric · Blockchain implementation

1 Introduction

The SALUS product quality and safety control system, offers innovative solutions on many levels. The analysis of the solutions available on the market showed that they are insufficient in view of the growing requirements resulting from the transfer of trade to the Internet and the specific requirements presented by the program's client. Existing solutions commonly use manual and/or direct methods of obtaining consumer opinions, often separated from the rest of the system, e.g. in the form of surveys conducted by pollsters. This is associated with high time-consuming, high cost and limited scope of research. The main feature of the IT system for product quality and safety control proposed as part of the project is a high level of automation of processes and functions, thanks to the use of AI mechanisms and blockchain technology [15–18].

On the today's market, customers have access to a variety of products of different brands. Many of these products are available in multiple countries, which is not surprising in a globalized market. However, products that bear the same or similar packaging and coming from the same brand are not always the same in terms of a composition, from which they were build. In research conducted in 2018/2019 European Joint Research Centre found that around 31% of the tested products had same or similar packaging but a different composition, and in turn different quality depending on a location [1]. This is what is called double quality.

Not surprisingly, customers desire products that have highest affordable quality. To obtain such products, knowledgeable potential customers research before making purchases. This research often consists of reading opinions, reviews of different products and comparing them. This is due to a fact that most customers do not have necessary knowledge to compare goods based on their composition, and thus must rely on experience of others. However, this issue could be solved by examining various products by dedicated laboratories and making results more accessible and available to a general public. Furthermore, customers should be able to influence which specific goods are being investigated.

In order to provide a solution, a system called SALUS is proposed. It is a public product quality and control system. It is intended to collect customers' opinions and reviews about various types of products and used them to identify goods that should be investigated. Results of this research would be also published in the system, so its users would be able to quickly check reviews and laboratory results of available products, thus detecting and mitigating issue of double quality.

SALUS fulfils these tasks by employing three modules: SALUS.OPN, SLAUS.BC and SALUS.AI. The first one is an UI module responsible collecting products' reviews, managing functionalities and for providing interface for users to interact with the whole system. The second one is blockchain network that is used as immutable database distributed between trusted entities to ensure system's credibility and more democratic access to its resources. Finally, the third one is a module utilizing machine learning, for classification of reviews and selecting products for laboratory investigation.

The contribution of the work described in the article includes the proposal of architecture for systems of this type, which are based on large amounts of data, which can be both static and dynamic, and use complex artificial intelligence and machine learning algorithms to process this data. The second important element of the proposed solution is the combination of the AI concept and Blockchain technology to create an efficient and safe system that processes large amounts of data, both homogeneous and heterogeneous, in real time.

A market review showed that such systems do not exist, especially in the open access version. SALUS system was created for the needs of a government agency and was intended to respond to the needs of society and public administration.

In this paper, SALUS.BC module is presented in detail. It is a backbone of the whole SALUS system. Blockchain technology provides the system with a secure and decentralized environment, which supports sharing resources and transparency due to its distributed nature. Furthermore, usage of smart contracts prevents unauthorized and illegal use of the system's resources.

The paper is structured as follows: Sect. 2 presents background of blockchain technology, Sect. 3 illustrates a general idea behind the SALUS system, Sect. 4 describes an architecture model od SALUS.BC module while Sect. 5 outlines implementation of SALUS.BC module with focus give to underling Hyperledger Fabric blockchain and provided be the module's API.

2 Blockchain Technology

Blockchain technology was introduced in 2008 by an entity called Satoshi Nakamoto [2]. Since then, it has been actively developed and employed in a wide variety of fields [3]. The main rationale behind Bitcoin was a need to provide a reliable, transparent, and trustworthy method of exchanging goods in a trustless environment that does not have supervision of third parties. In general, blockchain consists of a distributed and decentralized network of nodes, which work in a peer-to-peer environment. Each node maintains a copy of network-wide append-only ledger of transactions, which in order to be accepted into the data structure must first be collected into a block-like structure that must be verified and validated by a dedicated procedure. If the procedure is successful, such block is added to a ledger [4,5].

Today there are many implementations and distributions of blockchain technology. From a high-level perspective, they are differentiated by two criteria, namely: privacy and permission. Privacy characterizes whether a given blockchain is public or private, which relates to read rights of users. Specifically, public blockchain allows everyone to access data stored inside, while private blockchains limit access to a specific set of users. On the other hand, permissions refer to write rights, which characterize who can participate in verification and validation of transactions and append new blocks to network-wide ledger. In this case, blockchain can be permissioned, which means only a selected few can participate in verification and validation procedure, or permissionless that means everyone can do it [6].

As was already mentioned, there are multiple implementations and distributions of blockchain technology, so it would be impossible to describe them all or even majority of them. However, some of them are more well known and popular and for these reasons they will be briefly described. The first one is Bitcoin, which can be considered as a quintessential blockchain. It is a public and permissioned blockchain for managing cryptocurrency without external supervision. Procedure for verification and validation of blocks, i.e., consensus algorithm is represented by a Proof-of-Work algorithm. It is based on idea of hashing and a mathematical puzzle. In order to add a new block, nodes must compete in solving the puzzle. The first one to do so is allowed to add a new block to a ledger and receive cryptocurrency as a reward [2,5].

The second one well known distribution of blockchain technology is Ethereum. Similarly, to Bitcoin it is a public and permissionless network that manages accounts and cryptocurrency. However, Ethereum is a whole platform for various different applications that utilize blockchain and smart contracts that are implemented on top-of it. For consensus algorithm, Ethereum utilizes Proof-of-Stake, which selects validators based on proportion of their owned cryptocurrency [5,7].

The third and final distribution of blockchain technology is Hyperledger Fabric. Contrary to the previous two blockchain implementations, this one is private and permissioned, which means that it heavily restricts, who can access data and who can participate in a consensus algorithm. Hyperledger Fabric is another blockchain platform for developing distributed systems, but unlike Bitcoin and Ethereum, it is designed for more constricted environments like consortiums. Hyperledger Fabric has a modular architecture and is highly customizable, however by default it uses Raft algorithm for consensus It is based on a "leader and follower" model, in which a single electable leader is elected, and it is responsible for making decisions that are replicated by followers [5,8].

Blockchain technology is applied in many different fields, some of the most prominent ones include governance, financial sector, healthcare, education, and data management [3,5]. In governance blockchain is used to provide secure and transparent administrative services [9] or even electronic voting [6]. In financial sector main focus is given to cryptocurrencies [2,7]. On the other hand, blockchain in healthcare can provide platform for secure storage of records [10] and for managing telemetric sessions [11]. In education the technology can be used for storing academic data in a safe and secure way [12]. Finally, blockchain can be used in data management for sharing and monetizing data [13] or management of database metadata [14].

3 Idea and Scope of the SALUS System

The main idea of SALUS is to leverage blockchain technology and Artificial Intelligence (AI) in order to mitigate a risk of double quality and provide customers with information about quality of products on European market. To realize this goal the following functional requirements were specified:

1. Collect opinions about various categories of everyday products.
   (a) Opinions may be in one of three languages, i.e., Polish, English, and German.
   (b) Opinions may be in on of three formats, i.e., text, audio, and video.
   (c) Opinions must refer to products available on European market.
2. Analyse collected opinions to select products for examination in dedicated laboratories.
3. Store and share obtained information about product quality.

In order to fulfil these requirements, it was decided that SALUS needs to provide the following software elements:

1. Data Acquisition Component – for collecting product data.
2. Graphical User Interface – for interacting with the system.
3. Blockchain network – for immutable data storage with smart contract functionality for managing data sharing.
4. Artificial Intelligence Component – for analysing products in context of laboratory examination.
5. Product Quality Database – for audio and video type data and for mutable data required by Artificial Intelligence Component.

A general architecture and relationships between elements of the SALUS system are presented in Fig. 1.

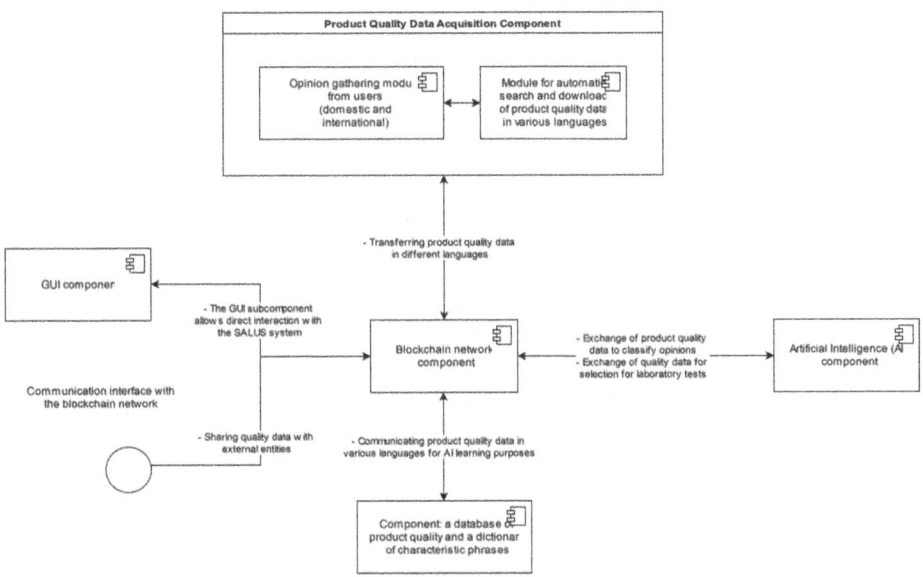

Fig. 1. Initial Architecture of the SALUS system.

The final version of SALUS system realizes these requirements through three integrated modules: SALUS.OPN, SALUS.BC and SALUS.AI. The first module is functions as User Interface (UI) for managing the system and for presenting and collecting data about available products. The second module is a backbone of the whole system, and its main task is to provide immutable and trustworthy data storage. Finally, the third module is responsible for analysing collected data and making decision about, which products should be examined and by which laboratories.

The purpose of SALUS.BC module is to provide a central database that stores data in a secure manner and ensures resistance to unauthorized modifications. The second purpose is to provide interfaces enabling integration and data exchange with other modules of SALUS system and with external entities joining the system. The main element of the module is a distributed and decentralized blockchain network based on Hyperledger Fabric distribution.

The purpose of SALUS.SI module to is the realization of following function with the use of selected machine learning methods:

- Classification of opinions in terms of sentiment.
- Product classification based on reviews.
- Analysis of text reviews to detect the double quality of products.
- Generating a report with the results of SALUS.SI component operation.

SALUS.OPN is the main application addressed to the recipients of the entire SALUS system. It is a web application that allows consumers to express their opinion on a given product, and for administrators and moderators to generate a report on the quality of a given product in terms of ratings.

This paper focuses on the SALUS.BC module and in the following sections a detailed description of the blockchain-based element of the SALUS system is presented.

4 Hyperledger Fabric

Hyperledger Fabric is an open-source implementation of blockchain technology provided by The Linux Foundation [8]. It offers an enterprise version of blockchain technology that relies on public and verified participants who reach consensus using the Raft algorithm. Hyperledger Fabric also offers support for smart contracts in Java, JavaScript and GoLang and an advanced logical security model. The main elements of Hyperledger Fabric are: (1) Organizations, (2) Peers, (3) Channels, (4) Chaincode, (5) Membership Service Provider.

Organizations that are the basic logical units that participate in the blockchain network. Organizations represent various business entities such as enterprises, institutions, government agencies, etc., which have their own unique identities and operate as independent entities in the network. Their basic function is to share and organize nodes (Peers) into logical entities.

Peers (Nodes) that are the basis of the blockchain network. Their role is to store copies of distributed registers (ledgers) and execute chaincode (smart contracts) in response to requested transactions. In a Fabric network, we can have different types of nodes such as Orderer, Peer, and Anchor Peer. Orderer (Ordering Service) is used to ensure consensus in the network and manage the order of transactions. The orderer is responsible for collecting transactions, organizing them into blocks and providing peer nodes with approved blocks that will be attached to local copies of the registers. Peers maintain ledger in a form of databases (by default LevelDB) and chaincode. Peers store blockchain in two forms: (i) World State that stores the current state of all resources (e.g. accounts, assets) on the network; (ii) Ledger/Blockchain that contains an immutable history of all transactions that have been approved. Finally, *Anchor Peer* are used for communication between organizations.

Channels are one of the elements that allow for privacy in Hyperledger Fabric. They allow the network to be divided into various private subnets. Each channel acts as a personal transaction space between selected participants, who may have different levels of access to data. *Chaincode* (smart contracts) are pieces of code that define the business logic and rules for processing transactions in the Fabric network. They can be written in various programming languages such as Go, JavaScript, or Java. *Membership Service Provider* (MSP) is responsible for authorizing users and verifying their identity. MSP allows you to determine who is allowed to participate in the network and also controls which certificates are accepted.

5 SALUS.BC Hyperledger Fabric Network

SALUS.BC is the central module of the SALUS product quality and safety control system. The main element of the module is a distributed and decentralized blockchain network based on the Hyperledger Fabric distribution [8]. The module consists of three basic organizations (in the sense of Hyperledger Fabric):

1. Org1 represents the SALUS.BC module.
2. Org2 represents the SALUS.SI module.
3. Org3 represents the SALUS.OPN module.
4. Org4 represents external entities that will be able to join the SALUS network.

The module supports a single channel dedicated to storing records for the module's resources: (1) Products, (2) Product Categories, (3) Review, (4) Revive Grade, (5) Reports, (6) Laboratories, (7) Standards.

The "Products" resource represents a set of products that are considered in SALUS and contains basic data about a given product. This includes product name, product manufacturer, product description, etc. The "Product Categories" represents different categories of products that are considered in SALUS, specifically: laundry detergent powders, dishwasher tablets, thermal mugs The "Reviews" is a resource that contains opinions of various types related to specific products. The "Review Grade" i s a resource that contains users "review of a review". In other words, it represents if other users found a given review helpful or not. The "Standards" resource represents a set of rules and regulations against which products can be verified. The "Laboratories" is a resource that contains data about various laboratories, which can investigate selected products. The "Reports" resource includes several different types of reports, the list and content of which may change in further iterations of the SALUS.BC module, depending on needs. In the current version, the "Report" resource includes two types, i.e., "Product Report" and "Rapex Comparison Report", the functionality of which is identical from the SALUS.BC perspective.

The acceptance of new blocks takes place through a simplified vote in which individual organizations take part. In the case of transactions carried out by internal organizations, the consent of at least one internal organization is required, i.e., one organization from the SALUS.SI, SALUS.BC and SALUS.OPN set. The same conditions apply to other operations that change the organization of the blockchain network. The last element of SALUS.BC is SALUS.BC API, which serves as an adapter that allows interaction with the system via a standardized interface, supporting smart contracts existing in the system (Fig. 2).

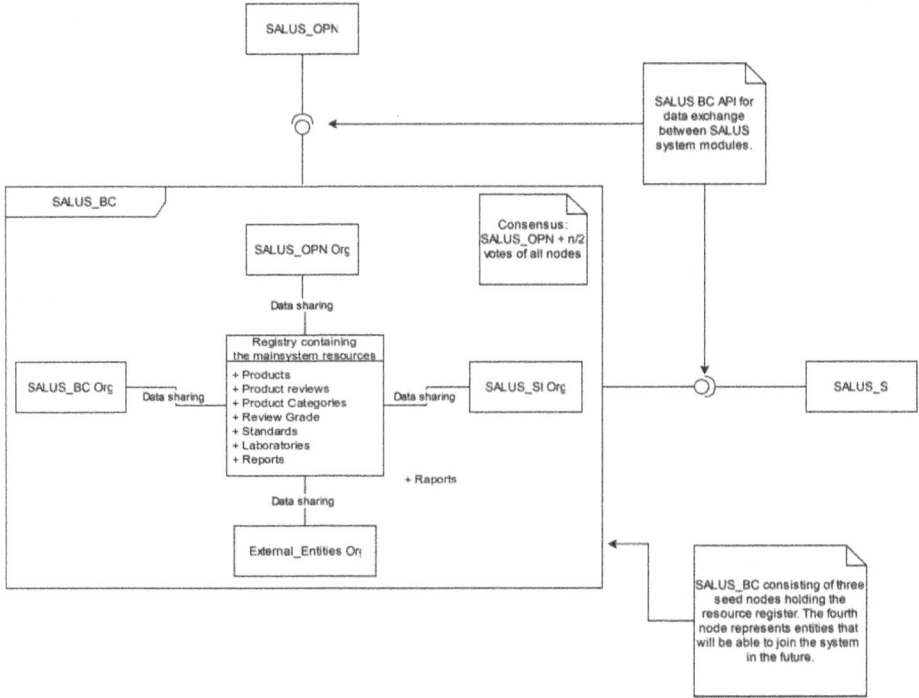

Fig. 2. Model of SALUS.BC in Hyperledger Fabric.

A very important part of SALUS.BC module is SALUS.BC API. It is designed and implemented in a REST architecture, which is universal and enables effective communication between servers, components and services. The assumption is that SALUS.BC will provide http methods to external components. Since blockchain is a data structure that only allows data to be added and not deleted or modified, the PUT, PATCH and DELETE methods are appropriately adapted to blockchain technology.

As part of the "Proof of Concept", each organization in the SALUS.BC Network module has a single node implementing all necessary operations. Orderer, on the other hand, has 3 nodes to increase network throughput. In total, this gives 7 nodes of the SALUS.BC Network module (4 organization nodes plus 3 Orderer nodes). Each organization node consists of the following elements:

– A set of smart contracts for the resources "Product", "Product Category", "Review", "Review Grade", "Report", "Laboratory" and "Standard".

- Ledger Instance.
- MSP (Membership Service Provider) instance.
- Instance of the SALUS.BC API component.

The elements of the SALUS.BC component communicate using a single "SALUS-1" channel. New blocks are added to the Registry through one of the three available Orderer nodes, which requires the consent of at least one of the internal organizations, i.e. SALUS.BC, SALUS.SI or SALUS.OPN. In Hyperledger Fabric notation: $OutOf(1, [SALUS.OPN, SALUS.SI, SAULUS_BC])$.

The Registry itself containing the data is built into the CouchDB database, which enables advanced queries. Figure 3 shows the SALUS.BC network diagram.

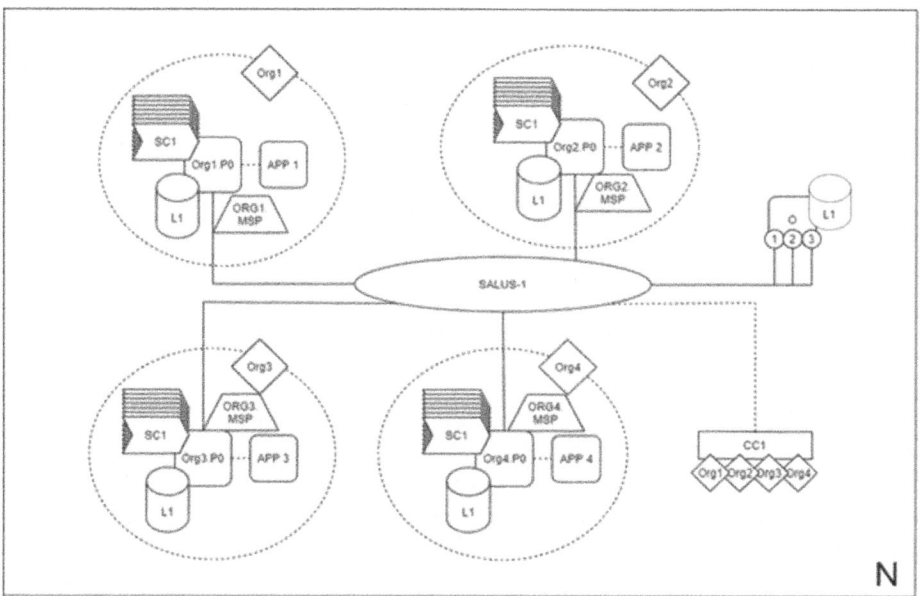

Fig. 3. Schema of blockchain network implemented in SALUS.BC network component.

Each resource in the SALUS.BC module is implemented using smart contracts, referred to as "chaincode" in Hyperledger Fabric. Figure 4 illustrates details of these resources and how they are constructed in a form of class diagram. Bold fields indicate "required" fields that always must be not empty.

Each chaincode provides methods for creating, downloading, updating, deleting and querying the resource to which a given contract applies. Figures 5 and 6 show the chaincode for all required resources and a list of methods that a given contract implements – it shows the smart contracts for the "Product" and "Product Category" resources, for the "Product Report" and "RAPEX Comparison Report" resources, for the "Opinion" and "Opinion Rating" resources, for the "Laboratory" and "Standard".

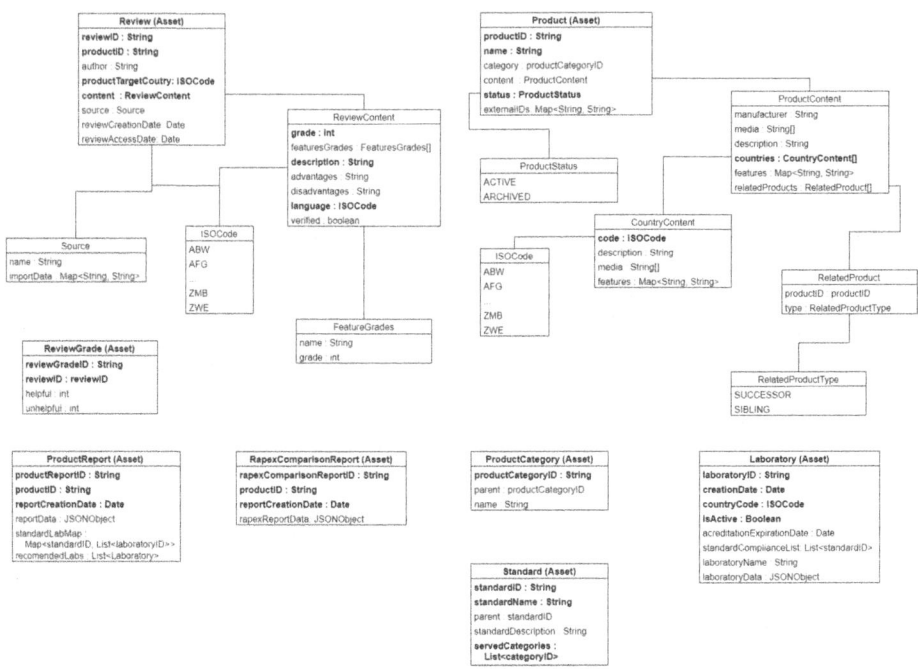

Fig. 4. Resources in SALUS.BC network.

The SALUS.BC Network module uses the Hyperledger Fabric infrastructure and can be easily and smoothly operated thanks to the open-source tool called Fablo. Fablo is a simple (but powerful) tool that makes it easier to work with Hyperledger Fabric, in particular prototyping, experimenting, and creating smart contracts. Within CI/CD pipelines, it also enables the automation of test cases and implementation processes. The repository contains all the necessary files to operate the network that has been configured with Fablo, as well as all the necessary smart contracts to control the business logic of the network. The SALUS.BC API is intended for use, which has been specially prepared to work with the network. Fablo is a simple tool for generating a Hyperledger Fabric blockchain network and running it in Docker. Supports RAFT and solo consensus protocols, multiple organizations, channels, chaincode installation and updating.

Fablo is used as a development and experimentation tool. This makes working with Hyperledger Fabric more convenient and simpler, reducing the cost of developing the SALUS.BC Network module in the PoC version. The repository contains two Fablo configuration files, one for development and one for the final version:

1. *fablo-config.json* – this is the default and development network, highly recommended when creating smart contracts using the Fablo REST API tool.

2. *fablo-config-final.json* – this is the network in the initial state of the proposed solution as the Salus.BC module.

With these files Fablo can configure the network, but additional actions may be necessary before the network is launched (e.g., changing the API image, changing names, connections, or longer sleep). Fablo is also used in the CI/CD environment. There are more complex and therefore better tools to achieve similar results, but this is best for local development.

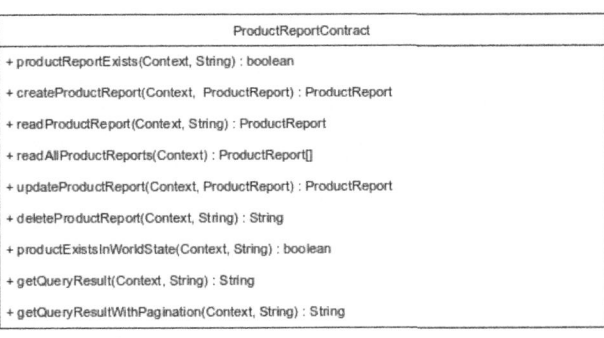

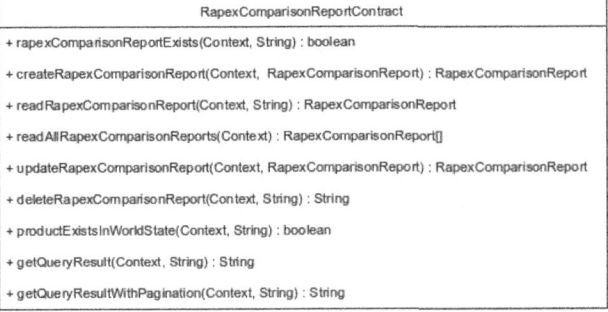

Fig. 5. SALUS.BC Product Reports and Rapex Comparison Reports chaincode.

6 SALUS.BC API

Each node of the organization has its own instance of SALUS.BC.API, which allows interaction with the interface it implements and interaction with the blockchain. The component is responsible for communication between the SALUS.BC Network component and entities external to the SALUS.BC module. SALUS.BC API enables CRUD operations to be performed for all resources specified in the functional specification. All functions provided by the API correspond to functions implemented in smart contracts (chaincodes) for resources in the SALUS.BC Network module.

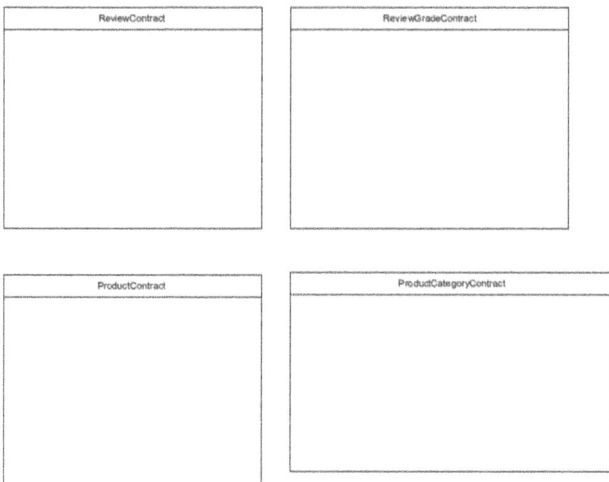

Fig. 6. SALUS.BC chaincode for Products, Product Categories, Reviews and Review-Grades.

First aspect of SALUS.BC API is authentication. Each endpoint except "/auth/enroll" requires a presence of an authentication token (Bearer Token) in the request. To generate a new token, it is necessary to register on the network, which can be done using the "/auth/enroll" endpoint. An example query can be found below. The default user credentials are defined by SALUS.BC Network.

```
curl --location --request POST '{address}/auth/enroll' \
--header 'Accept: application/json' \
--header 'Content-Type: application/json' \
--data-raw '{
"id": "secret-id",
"secret": "secret-password"
}'
```

Assuming the credentials provided are correct, a response similar to the one below can be expected to be returned. Please note that a token is always a randomly generated hexadecimal string of 128 characters long. The token is valid for one hour from the time it is generated.

```
"content": {
"token": "b139228c2...afd0f173347"
},
"statusCode": 200
```

Each index endpoint (i.e. [GET] /channel/asset) supports queries. To query data, the fields and their desired value must be provided as an HTTP query in the following format:

```
%[GET] /{channel}/{asset}?queriedField=desiredValue\&anotherQueriedField=
otherDesiredValue
```

In the case of multiple queries (as in the example above), the system will use an AND operation to combine them, i.e., in the above example, only those resources that have "queriedField" set to "desiredValue" AND "anotherQueriedField" set to "otherDesiredValue" will be returned.

A simplified example of such a resource is provided below. Note that both the query field and the requested values are case-sensitive and exact values are needed, e.g., "productname" does not match "productname".

```
"queriedField": "desiredValue",
"anotherQueriedField": "otherDesiredValue",
"irrelevantField": "irrelevantValue"
```

To query for nested parameters, dot notation must be used as shown below with a simplified example of matching a resource:

```
%[GET] /{channel}/{asset}?queriedObject.queriedArray.queriedField=
        desiredValue
```

```
"queriedObject": {
"queriedArray": [
{   "queriedField": "desiredValue"   },
{   "queriedField": "otherValue"     }    ],
"irrelevantField": "irrelevantValue"
   }
```

Below is a practical example of a query for products with a defined name and products with the word "Hello" in the description of the country of occurrence:

```
[GET] /{channel}/products?name=desiredProductName
[GET] /{channel}/products?content.countries.code=POL\&content.countries.
description=Hello
```

Each endpoint index supports pagination. To split data into pages, the pageSize field must be provided in the HTTP query string. The pageSize value indicates the desired number of records per page in the response.

```
%[GET] /{channel}/{asset}?pageSize=15
```

When pagination is used, additional field metadata will be sent in the response, this field contains a bookmark field, which is a string that can be used to retrieve records from the next page – as shown below (bookmark must be replaced with the value returned in the previous request):

```
[GET] /{channel}/{asset}?pageSize=15\&bookmark={bookmark}
```

Pagination works with and without query operation. Below are examples of how to get five labs per page and how to use a query to get fifteen products per page that are sold in Poland.

```
[GET] /{channel}/laboratories?pageSize=5
[GET] /{channel}/products?content.countries.code=POL\&pageSize=15
```

Each successful response (HTTP code 2XX) has the format shown below, where the content field contains the actual content of the response (for example, product details) and the statusCode field represents the HTTP status code.

```
"content": { ... },
"statusCode": 2XX
```

Each failed response (HTTP code 4XX or 5XX) has the format shown below, where errorType is the error name stored in CONSTANT_CASE, the content field contains the actual content of the response (for example, error details), internalErrorCode, which is the internal error code of SALUS.BC Network, or SALUS.BC API (described below) and a statusCode field representing the HTTP status code.

```
"errorType": "ERROR\_TYPE\_NAME",
"content": { ... },
"internalErrorCode": "XXXXXX",
"statusCode": 4XX
```

7 Role SALUS.BC in the SALUS System

SALUS.BC is a backbone of the whole SALUS system. It provides distributed storage based on blockcahin system. The current PoC is centralized, in a sense that all organizations are controlled by a single entity. However, the intended final state will distribute control between independent organizations and institutions, which will ensure that data is immutable and not modified to suit someone needs. Thus, for the release the SALUS.BC Network will be configured in such a way that there initially are at least 3 independent organizations participating in RAFT consensus algorithm.

SALUS.BC provides "backend" for the system. Smart contracts provide CRUD functions for data stored in blockchain and API provides more user/developer friendly interface for using these smart contracts. Furthermore, SALUS.BC provides part of access control for SALUS. Each function, be it from smart contracts or API is protected and requires proper authorization, so no unauthenticated entity can use them. Utilization of smart contacts in Hyperledger Fabric and API provides logging functionality through which all action on the system is recorded. Thus SALUS.BC provides monitoring and control functionalities for SALUS.

In a context of other modules od SALUS, SALUS.BC provides them with data source for data that should not be mutable and is crucial for building trust in the system. What it entails is covered in Fig. 6.

8 Conclusions

SALUS.BC is the central module of the SALUS product quality and safety control system. The main element of the module is a distributed, decentralized and private blockchain network based on one Hyperledger Fabric. The use of a private blockchain network allows to implement access control and limit interference in the system to trusted and verified entities. This also makes it possible to define a consensus algorithm suitable for a non-public blockchain network, e.g., democratic voting. It is worth mentioning that the "non-public" blockchain network means that access to data and joining the system as a full-fledged node is limited by an access control model consistent with a previously established policy consistent with the interests of stakeholders. This does not mean centralization of the system and confidentiality of data. The use of a non-public network allows both to increase system security and to reduce maintenance and usage costs, as it reduces the need for resources associated with expensive contention-based consensus algorithms, e.g., proof-of-work.

The combination of blockchain technology and selected AI mechanisms within one software is an innovative solution. The review of existing solutions showed that currently there are not many systems on the market that combine these two technologies for practical purposes. Thus, the system proposed in the project, combining the blockchain network, distributed database, and Artificial Intelligence (AI) performing tasks related to data processing and analysis, is an innovative approach to solving the presented problem of double quality of products. This provides the software with a wide range of functions associated with a high level of automation and an extensive data security policy. An innovative element of combining AI with blockchain technology is to ensure transparency and testability of the entire system. Typical systems based on AI do not allow for detailed tracking of all operations carried out. Thanks to blockchain technology, it is possible to save the results of AI methods as part of transactions in network blocks, which allow for the verification and validation of operations carried out by the system, based on reliable and unmodifiable data.

The SALUS.BC component was integrated into the SALUS system with success. The effectiveness of the proposed system were evaluated according to the following criteria:

1. Automated classification of opinions into positive, negative and neutral – minimum 91% of the correct classification of the collected opinions.
2. Support for different feedback formats – numeric, text, audio and video.
3. Automated selection of products for laboratory tests – minimum 91% of correct selections.
4. Support for different languages – the first version of the system supports the processing the opinions in Polish, English, and German.
5. Secure sharing of data with external entities – data sharing interface using a secured blockchain network.
6. Verifiability of the functionality of AI module – recording operations in the form of irreplaceable transactions in the blockchain network.

Acknowledge. The presented work was supported under the grant of Polish National Center of Research and Development, Project No. INFOSTRATEG-III/0004/2021-00.

References

1. Joint Research Centre: Same pack, different ingredients? Dual quality down in branded EU food (2023). https://joint-research-centre.ec.europa.eu/jrc-news-and-updates/same-pack-different-ingredients-dual-quality-down-branded-eu-food-2023-07-24_en
2. Nakamoto, S.: Bitcoin: A Peer-to-Peer Electronic Cash System (2008)

3. Casino, F., Dasaklis, T., Patsakis, C.: A systematic literature review of blockchain-based applications: current status, classification and open issues. In: Telematics and Informatics, vol. 36, pp. 55–81 (2019)
4. Drescher, D.: Blockchain Basics. A Non-Technical Introduction in 25 Steps. Apress, Berkeley, CA (2017). https://doi.org/10.1007/978-1-4842-2604-9
5. Singhal, B., Dhameja, G., Panda, P.S.: Beginning Blockchain. Apress, Berkeley, CA (2018). https://doi.org/10.1007/978-1-4842-3444-0
6. Pawlak, M., Poniszewska-Maranda, A.: Trends in blockchain-based electronic voting systems. Inf. Process. Manage. **58**(4), 102595 (2021)
7. Wackerow, P.: Ethereum Development Documentation. In: Ethereum (2022). https://ethereum.org/en/developers/docs/. Accessed 03 Dec 2023
8. Hyperledger Foundation: A Blockchain Platform for the Enterprise. In: Hyperledger Foundation (2022). https://hyperledger-fabric.readthedocs.io/en/release-2.5/. Accessed 03 Dec 2023
9. e-Estonia. Frequently asked questions estonian blockchain technology (2020). https://e-estonia.com/wp-content/uploads/2020mar-nochanges-faq-a4-v03-blockchain-1-1.pdf. Accessed 03 Dec 2023
10. Ekblaw, A.: MedRec: blockchain for medical data access, permission management and trend analysis. Doctoral dissertation, Massachusetts Institute of Technology, Cambridge (2017)
11. Medicalchain SA: Medicalchain Whitepaper (2017). https://medicalchain.com/Medicalchain-Whitepaper-EN.pdf. Accessed 03 Dec 2023
12. Medina, J.: Educational records and the blockchain (2017). https://www.learningmachine.com/wp-content/uploads/2017/03/BlockchainforEducation-1.pdf. Accessed 03 Dec 2023
13. Ocean Protocol Foundation Ltd: Data: The New Asset Class – Ocean Protocol. In: Ocean Protocol Foundation Ltd, (2023). https://oceanprotocol.com/about-us/ocean-token. Accessed 03 Dec 2023
14. García-Barriocanal, E., Sánchez-Alonso, S., Sicilia, M.-A.: Deploying metadata on blockchain technologies. In: Garoufallou, E., Virkus, S., Siatri, R., Koutsomiha, D. (eds.) MTSR 2017. CCIS, vol. 755, pp. 38–49. Springer, Cham (2017). https://doi.org/10.1007/978-3-319-70863-8_4
15. Dave, K., Lawrence, S., Pennock, D.M.: Mining the peanut gallery: opinion extraction and semantic classification of product reviews. In: Proceedings of 12th International Conference on World Wide Web, WWW 2003, pp. 519–528 (2003)
16. Li, J., Kassem, M.: Applications of distributed ledger technology (DLT) and blockchain-enabled smart contracts in construction. Autom. Constr. **132**, 103955103955 (2021)
17. Zhang, R., Xue, R., Liu, L.: Security and privacy on blockchain. ACM Comput. Surv. **52**, 1–34 (2020)
18. Sherman, A.T., Javani, F., Zhang, H., Golaszewski, E.: On the origins and variations of blockchain technologies. IEEE Secur. Priv. **17**(1), 72–77 (2019)

QUALIFIER Workshop

Mapping Source Code to Software Architecture by Leveraging Large Language Models

Nils Johansson[1,2](✉)[iD], Mauro Caporuscio[1][iD], and Tobias Olsson[1][iD]

[1] Linneaus University, Växjö, Sweden
mauro.caporuscio@lnu.se, tobias.olsson@lnu.se
[2] Volvo Construction Equipment, Braås, Sweden
nils.johansson@lnu.se

Abstract. Architecture refactoring is a big challenge and requires thorough analysis and labor-intensive, error-prone activities to restructure functionalities from a legacy architecture to a new intended one. Indeed, source code should be adapted to match the new structure. In this context, automatically mapping source code to the intended architecture would significantly reduce manual work and prevent technical debt. To this end, in this paper, we aim to map *methods* to architectural modules solely defined by textual descriptions, i.e., formulated as a machine learning text classification problem. Methods are mapped into modules using different approaches. We apply the proposed approach to an open-source software system, results show that vectorizing text and code using large language models outperforms other modern methods. The different applied machine learning classifiers perform comparably well, where the best attain accuracy of around 40% and F1-score of around 30%.

Keywords: software architecture · software refactoring · source code mapping to architecture · large language models · machine learning

1 Introduction

Software development organisations apply software refactoring to improve quality aspects of software systems. Refactoring changes the implementation of software without altering its behavior, i.e., the functionality from a black-box perspective is unchanged.

Refactoring of software may be done at both code and architecture levels. In *Code Refactoring* updates are applied locally and do not affect the structure and interrelationships of modules. On the other hand, *Architecture Refactoring* aims to change the software architecture, reorganizing the existing code into a new structure. A major challenge in architecture refactoring is trying to reuse as much existing code as possible to avoid writing new software.

While several tools and methodologies exist to facilitate, and even automatize, code refactoring [5,7,10], similar tools for architecture refactoring are not as common. Recently, some methods based on machine learning demonstrated to be effective and able to predict and recommend refactoring possibilities by training on a vast number of software projects [6,9,21].

Nevertheless, if a software organization is in a situation where they have a new, intended architecture, e.g., based on a reference architecture or similar, these tools are not applicable. A tool where a reference model is input and a proposed new software system is determined from the existing code is not available today. This study addresses the following use case: given a software system implemented according to a legacy architecture and a new intended architecture defined as textual descriptions. Along with diagrams, textual documentation is commonly to describe an architecture [27]. We seek a mapping between code entities (e.g., code line/lines, methods, blocks, or classes) from the old architecture to the new architecture (where it should be placed). After unsuccessfully finding a suitable refactored software system where the mappings of old code to a new intended architecture can be traced, we decide to try to map code entities to its existing module. Such a mappings would be valuable for software architects and developers in the refactoring process.

The problem is approached as a machine learning classification method, where descriptions of modules are used to train a classifier. The classifier is evaluated by predicting the module label of methods found in the source code and comparing to the ground truth. We investigate different methods for preprocessing the module descriptions and for vectorizing the descriptions and code using techniques from the Deep Learning field, including Large Language Models (LLMs). Results indicate that using LLMs to vectorize text outperform other methodologies in terms of classification accuracy. However, the results do not favor any specific classification technique over the others.

The paper is organized as follows. Section 2 formulates the problem addressed by this work, whereas Sect. 3 discusses the related state of the art. Section 4 describes the proposed approach and the set of experiments we designed to evaluate it. Obtained results are reported in Sect. 5 and analyzed in Sect. 6. Finally, Sect. 7 concludes the paper and provides hints for the future.

2 Problem Formulation

Let $SourceSet = \{c_1, c_2, \dots\}$ be a set containing a number of code blocks c_i, which may include several lines of source code and originate from an arbitrary number of files. Let $TargetSet = \{m_1, m_2, \dots\}$ be a collection of architectural modules m_j, each containing a textual description of the module itself. Every code block c_i belong to one target architectural module m_j, i.e., the code block is or should be implemented in one module, depending on if the code follows the intended architecture. Therefore, we seek a function $Mapping : SourceSet \rightarrow TargetSet$ that, given a code block c_i, returns the architectural module m_j to which c_i should belong (see Fig. 1).

Mapping of source code to architectural modules may be formulated as a machine learning classification problem. In such problems, one tries to identify

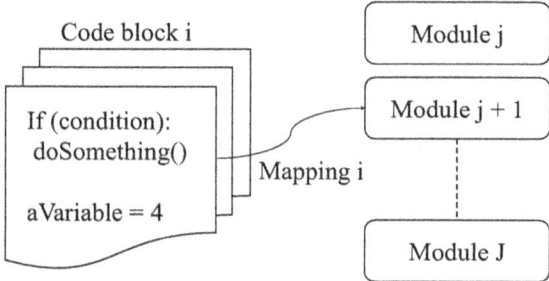

Fig. 1. Mapping between source code blocks and architectural modules.

the class of a data point using a number of input *features*. Beforehand, a classifier is trained on a labeled dataset, denoted as training data. One can then evaluate its accuracy by predicting the class on a set of testing data points and comparing with the ground truth. In our setting, we want to predict the class of source code entities using training data from descriptions of modules. When using text data, a required step is vectorization, where text is translated to real-numbered vectors. By vectorizing text data using different approaches, including LLMs, one can compare the result of applying these methods. This study attempts to answer the following research questions:

RQ1 How do the classification accuracy compare between different methods when classifying source code to modules described in natural language?

RQ2 What factors inhibit an accurate result when classifying source code to architectural modules?

3 Related Work

There exist several methods to map source code entities to architectural modules, both manual or with elements of automation. These have most often been developed for the purpose of being used in software architecture consistency checking or reflection modeling.

An early approach that featured automation is the method "Human Guided" clustering method [12], which relies on an initial mapping between source code files and architectural modules made by a practitioner. Thereafter, an algorithm is used for automatically processing the remaining source code entities not yet mapped.

Later, Bittencourt et al. [11] proposed using semantic analysis of text, which performed as well as the "Human Guided" clustering method. A different approach, called RELAX [23], suggests using any type of textual documentation of the modules. This documentation is used to train a machine learning classifier that classifies source code files into architectural modules. The purpose of this approach is to perform architectural recovery, i.e., extracting valuable information from the source code of an existing software system.

The performance of these methods is primarily influenced by the quality and size of the initial mappings made by a human practitioner, which may deter users from applying these techniques. To overcome this limitation, Herold and Sinkala [32] showed how to exploit the textual descriptions of software modules that often exist in documentation to recommend the initial mappings. Descriptions are processed to compute *similarity scoring* between modules and source code. Similarities between source code and architectural modules may be described in several different ways. *Structural similarity* is the similarity of dependencies between components. This is not straightforward to use as other similarities when mapping source code for refactoring since the structure between the implemented code and intended architecture might differ [15]. Other similarity metrics include *linguistic* and *concern* similarity. Entities are linguistically similar if words are equal or, in some sense, close to each other. Source code and modules can be similar regarding their concerns if these are shared between them, such as dealing with interaction or logic functionality. Florean and Jalal [17] evaluate two different methods, namely Bag-of-Words (BoW) and inverse document frequency (TF-IDF), to compare similarities between words and found that BoW performs slightly better. Using either of these methods requires that the same terminology is used for both data sources, i.e., module documentation and source code. Olsson et al. [26] introduced a method which leverages machine learning to classify source code entities. Specifically, the approach relies on an initial set of seed mappings and then calculates an attraction function based on Naive Bayes. Their approach consists of aggregating information found in the source code from multiple sources based on relationships between files, which improves the resultant mappings. Similarities between code entities and modules is calculated using BoW. Florean et al. [18] evaluated three different machine learning classifiers and the extent to which seed mappings affected performance. They found that *Logistic Regression* and *Support Vector Machines* perform better than Naive Bayes in mapping source code entities.

Several of the methods previously established rely on finding similarities between words used in the code and architectural module documentation. This may pose a problem if, e.g., different people use different terminology to convey the same message. Furthermore, the mapping is usually done on a file-level, meaning that individual source code files are mapped to modules. Certain refactoring activities may require splitting files or, in some way, restructuring modules using different parts of multiple files. Mapping with finer granularity, such as a set of lines of code, may enable the technique to be used for additional purposes than software architecture consistency checking. This can negatively affect the classification, especially when applying a BoW or TF-IDF vectorization, since it requires an overlap of exact terminology [16] of the texts that are compared. Comparing text with code is increasingly difficult as the code segment is smaller, and there is less information in the code to use. To address this, one can apply modern techniques from machine learning-based text classification problems. Firstly, there are available *text embeddings* that are trained to translate text to vectors based on similarity. Words that are similar will be translated to ndimensioned vectors of similar values. Examples of such embeddings are Global Representation Vectors (GloVe) and FastText. Secondly, Large Language Models are deep neural networks trained for similar purposes.

Popular LLMs used in text classification include the Bidirectional Encoder Representations [14], and its many variants (e.g., [13]). One can utilize both general-purpose pre-trained models and domain-specific models [13]. Models that are trained on code as well as natural language are available [25]. These models are developed for code-related tasks such as code -search [36], -summarization [38] or -generation [19]. These may be used e.g. to interpret and extract vectorized features from code, which can then be used in a classification algorithm.

4 Experiment Design

To evaluate the performance of different classification techniques including e.g. using pre-trained, large language models, we use the following approach, as shown in Fig. 2. The workflow consists of three major parts: **training**, **prediction**, and **evaluation**. Hereafter follows a brief description of the approach, and further details are described in Sects. 4.1 and 4.5. This process is repeated using varying components, such as vectorization methods, classifiers, and hyperparameters.

Training – Textual descriptions of architectural modules are generated using GPT 3.5. These are pre-processed and augmented using different methods. Thereafter, the text data points are vectorized and used with the module labels to train a machine-learning classifier.

Testing – The source code of the open source software system *PX4* is parsed. We choose to attempt to map *methods* to architectural modules and, therefore, extract these from the code during parsing. We also extract names of *user-defined identifiers* since they are deemed to carry significance with respect to architecture. This data is thereafter pre-processed and vectorized by the same method used for the training data. Next, we predict labels of *methods* using the vectorized text of either the whole method content or the identifiers.

Evaluation – Finally, the predicted labels are compared to the true labels. Accuracy and F1-score of the classifications are calculated.

4.1 Generation of Module Descriptions

To study how small code entities may be mapped to software architecture modules, we select the software system PX4. It is an open source [29] embedded software system for the control of flying drones, implemented in C++ [1]. However, detailed descriptions of the modules are missing for nearly all class definitions.

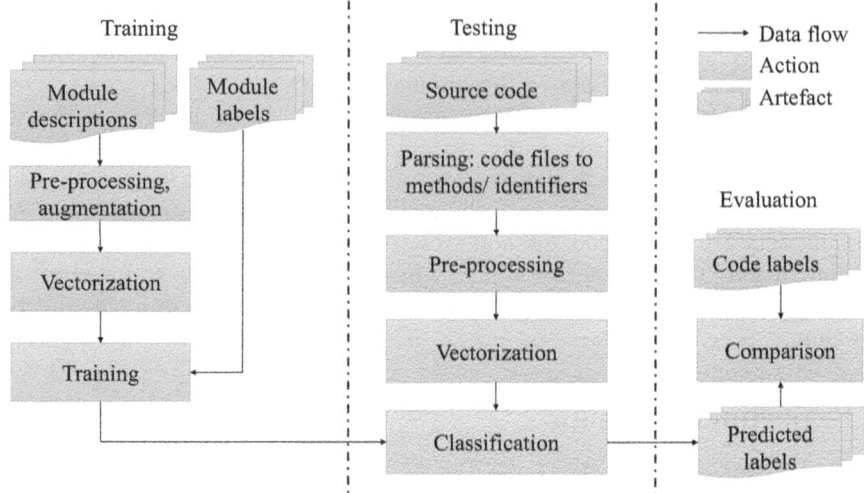

Fig. 2. Mapping approach.

Only a few classes are described in text, and the descriptions of modules on the highest level in the hierarchy are very short. Therefore, we choose to generate descriptions of the C++ class definitions using OpenAI's GPT 3.5. To decrease the risk of these descriptions being similar to the code [37] and severely introducing bias, this is done in three consecutive steps by three different prompts:

1. Translate c++ code, row-by-row, to natural language. Include all details such that the text matches the functionality of the code. It should be possible to reconstruct the code from the text. Ignore the initial large commented block.
2. Describe the functionality of the software component. The length of the response must range from 75 to 100 words. Avoid specific method mentions and focus on the overall purpose and functionality of the module. Focus on describing the role of the module in the context of other modules.
3. Paraphrase the following description of a software component. The length of the response must range from 75 to 100 words. Make it appear as if it has been written by someone else. Use synonyms for words found in the text and employ varied phrases such that the language is different and the wording is dissimilar but still conveys the same meaning.

The input of the first prompt is the code of a class, i.e., the contents of a ".cpp"-file, and the output is that code translated to natural language. That result is then used for the second step, where the translated code is described as a software component. Lastly, the description of a component is paraphrased to alter the language. The purpose is to make the text appear as if two different people implemented the code and wrote the description, e.g., a software developer and an architect. The class descriptions are then aggregated according to the module they belong to in the highest level of the source code hierarchy[1]. We accomplish this by labeling all training data as the module name. This level has 45 modules constituted by a total of 359 sub-modules, meaning that the classification problem has 45 classes and data from 359 descriptions, which is later used as the training data.

4.2 Parsing of Source Code

We select to map *methods* from source code to modules. The methods and their contents are extracted by a Python script that uses *regex* patterns to find methods in the source code. In addition, we extract all *user-defined identifiers* from these methods, which are artifacts such as variable and method names that have been defined by a human. These words are particularly interesting from an architectural perspective, since they are entities that results from design decisions. This parsing is shown in Fig. 3.

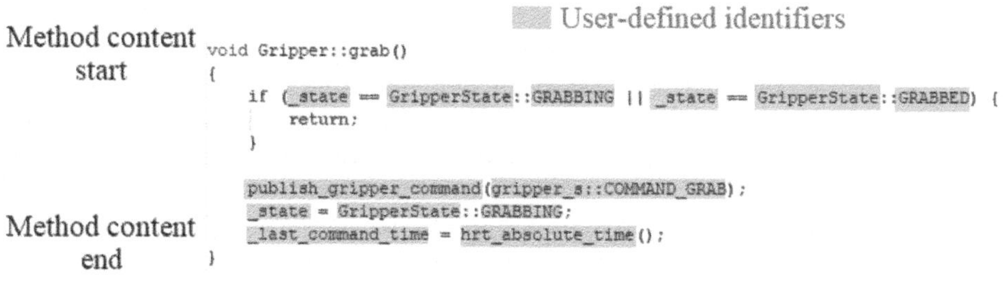

Fig. 3. Data extraction from a method.

4.3 Pre-Processing and Augmenting

Training Data. Before using the training data (i.e., module descriptions) to train a classifier and testing data (source code artifacts) for prediction, we apply a few variable pre-processing and augmentation techniques. Augmentation is performed by resampling from the original descriptions, as described in Table 1. The rationale for evaluating the use of different text formats is to see how the classification performs when using data points encompassing larger information amounts (e.g., whole paragraphs) versus smaller (e.g., sentences or snippets).

---

[1] see GitHub: "PX4/PX4-Autopilot/tree/main/src/modules".

Table 1. Processing of architecture module text descriptions.

Text Processing	Explanation
Paragraphs	Textual descriptions of one whole class are used as a single data point.
Sentences	Each sentence is represented as a data point.
Sliding window	Text snippets are extracted from the text using a sliding window algorithm. The sliding window width is chosen as either three or nine words or a combination. The sliding window moves one word to the right and selects a new sequence of words [22,34].

Next, the training data is filtered to remove all points where the feature values(i.e., the text) are identical but the label is not. In other words, if two or more different modules have the exact same sentence (or sequence of words when using a sliding window), the data points of these sentences are removed. This step removed many generic terms.

Testing Data. The identifiers from the test data are processed to more closely resemble natural language. These steps include removing identifiers that are only numbers or single letters and removing points where the feature data (i.e., the identifier name) is identical to the label. Since many identifier names were found to be equal, we choose to filter points where the feature value exists for more than one label value. For the remaining points, we converted the camel case to white spacing and replaced underscoring with white spacing. Lastly, we then convert all values to lowercase.

4.4 Vectorization of Training and Testing Data

Next follows the vectorization of the training and test data. Firstly, we choose to include a set of the common vectorization methods, TF-IDF, BoW, and the word embedding GloVe, which is implemented using Word2Vec [28]. We include two large language models in addition to the previous three vectorization methods. A well-performing model in general text classification problems "all-MiniLM-L6-v2" [2,3] implemented using the *SentenceTransformers* python library. Lastly, we select a domain-specific model that is trained on C++ code, called *UniXCoder-base-nine* [4,19]. We choose to use only pre-trained models, as training a new language model or fine-tuning a trained model, using only the available data from the module descriptions is not feasible. Even though after applying some of the data augmentations (e.g. using a small sliding window) we attain a larger data set, this is deemed not sufficient for a multi-class problem [8,20,33,35]. Furthermore, we choose not to evaluate zero/one/few-shot classification by prompting [30], as this would require including all training data in the prompt. Usually, generative AI applications such as GPT-3.5 limit the input prompt length. We then find shot classification unreasonable for this use-case, as a practitioner would always prefer to utilize all available data.

4.5 Classification and Evaluation

The vectors are thereafter used in training a set of different machine-learning classifiers. These are described in Table 2. Suitable hyperparameter values were selected after some initial trial tests. If unspecified, we used the default setting of the Scikit-learn library. The neural network used consists of one input, one output, and three hidden layers. The input and hidden layers uses a ReLU activation whereas the output a softmax function. Between each layer, we use a dropout rate of 0.5. The number of neurons in the input layer is forced to be the number of input dimensions. The size of the output layer is the number of classes, predicting the probability that the data point belongs to a specific class. For the hidden layers, we select to use half of the number of neurons of the previous layer [8,16,24].

Table 2. Classification algorithms used.

Algorithm	Hyperparameters
Gaussian Naive Bayes	Scikit-learn default settings
k-Nearest Neighbour	$k = 3, 5, 7$
Logistic Regression	tol = 0.01, max iter = 25
Support Vector Machine	tol = 0.01, max iter = 25, kernel = rbf
Shallow Feed-forward Neural Network	max epochs = 30

Finally, to predict the class of methods using the test data from the source, we employ a few different methods. Firstly, the *method content* is vectorized by the same model as the training data as described in Sect. 4.4. It is then labeled using one of the trained classifiers. The second approach of classifying the methods uses the *user-defined identifiers* found in the contents of a method. Each identifier is thereafter classified, meaning that it is possible that the identifiers are classified into different modules. Two different approaches are applied to classify the whole method. The first is simply selecting the most common class, i.e., voting. For the third approach, the highest summarized probabilities of all identifier predictions are used. If a method has n identifiers connected to it and there are c possible classes, predicting the probabilities of these yields n vectors of size c. The class is selected to be the class with the highest summarized value over the n vectors. This method is only used for the algorithms that use probabilities for prediction, i.e., Naive Bayes, Logistic Regression, and the Neural Network. The results are evaluated by calculating accuracy and F1-score [31]. The complete classification reports of all methods are saved for later analysis. Lastly, to evaluate generalizability, we perform a test where 20% of the training data is randomly removed from the set.

5 Results

The results of all configurations tested are shown in Fig. 4. It shows that augmenting the data using the sliding window technique and vectorizing by large language models generally provide the best result in terms of classification accuracy. Both language models *MiniLM* and *UniXCoder* perform better than other vectorization methods. As for classification, finding a trend is not as clear. Logistic regression, kNN, and the neural network perform best and are comparably accurate.

Table 3 includes the detailed result of the five configurations that yielded the highest accuracy and F1-score. The result of the test where a portion of training data was held out from training the classifier is described in Table 4. A slight decrease in accuracy and F1-score is attained when removing 20% of the training data.

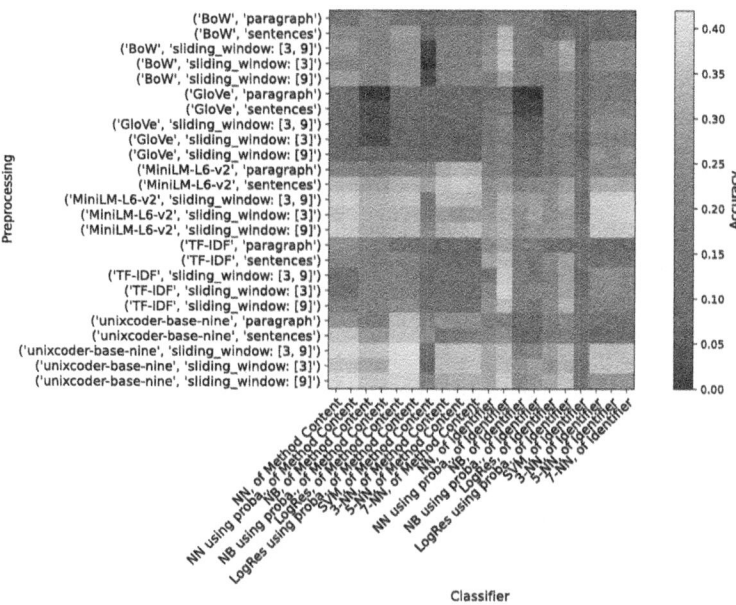

Fig. 4. Classification results with training data.

6 Analysis and Discussion

From the results, we see that using probabilities to label a method using identifiers often performs better than a voting approach. It is reasonable to assume that some identifier names are more difficult to label than others, e.g., if the software developer decided to use a less descriptive name. These should not weigh as heavily as other identifiers, which is what the results show. The dataset is very imbalanced, so we analyze the classification results of each class. In Fig. 5, the correct and incorrect predictions of each class are shown in relation to the amount of training data. The training data amount is shown as a grey hatched bar as the number of words in the data set. The outcome of the predictions is shown as stacked colored bars with their designated y-axis, where the color describes if a data point, i.e., a method, was correctly or incorrectly classified. In these bar plots, it is difficult to conclude that the imbalance between training data has affected the results significantly. Even the classes with more training data are often misclassified. However, there is one class, *ekf2*, that has a relatively medium amount of training data to which many test data points (methods) belong in truth to other classes but are labeled as this class. Additionally, there is a class *fw_pos_control* that has very little training data but a large number of test points, of which nearly all are misclassified. This means that this class

Table 3. Result of the highest performing configurations.

Data Creation method	Vectorization model	Evaluation data	Classifier	Accuracy	F1 Score
sliding window: [3,9]	unixcoder-base-nine	Method Content	Logistic Regression	**0.413**	0.244
sliding window: [3]	unixcoder-base-nine	Method Content	Logistic Regression	**0.403**	0.252
sliding window: [9]	unixcoder-base-nine	Method Content	Shallow NN	**0.398**	0.184
sliding window: [3,9]	sentence transformer	Identifier	3-NN	**0.396**	0.264
sliding window: [3]	tf-idf	Identifier	Shallow NN using probabilities	**0.396**	0.279
sliding window: [9]	sentence transformer	Identifier	3-NN	0.387	**0.290**
sliding window: [9]	sentence transformer	Identifier	5-NN	0.374	**0.279**
sliding window: [3]	tf-idf	Identifier	Shallow NN using probabilities	0.335	**0.279**
sliding window: [9]	sentence transformer	Identifier	7-NN	0.376	**0.276**
sliding window: [3,9]	sentence transformer	Identifier	7-NN	0.390	**0.268**

Table 4. Result of the training data hold-out test.

Used training data	Data creation method	Language model	Evaluation data	Classifier	Accuracy	F1 Score
100%	sentences	sentence transformer	Method Content	5-NN	0.388	0.204
80%	sentences	sentence transformer	Method Content	5-NN	0.369	0.179

consists of few files but has many methods in those files. With limited training data, the code of this class is likely difficult to predict.

Next, we continue to analyze how the relationship between the amount of code and the amount of training data affected the classification performance. This is of interest since the amount of training data generated was chosen as max 100 words per class. From Fig. 6, where code per file in lines of code is plotted against the amount of training data per class in a number of words as a scatter plot. Each point represents classified methods in a file, and its color shows the accuracy of the predicted labels of the methods, where an accuracy of 1 means that all methods in a file were correctly classified. One can see that points with high amount of training data but few LoC per file seem to be slightly better classified, but the trend is not clear. In machine learning, more data is generally advantageous, so it is not unreasonable to assume that it would hold in this case as well. However, other factors have likely affected the predictions as well, e.g. in this case that some modules/classes have more unique language and would then be easier to classify.

In machine learning classification problems, the availability of accurate training and test data is crucial. However, for this problem, the training data is constituted by textual descriptions of modules (in natural language), whereas the test data are artifacts of source code(programming language). This has almost certainly affected the results. Still, since the LLMs perform more accurately than other vectorization methods, these are likely better at finding true similarities between natural and programming languages. However, it should be noted that

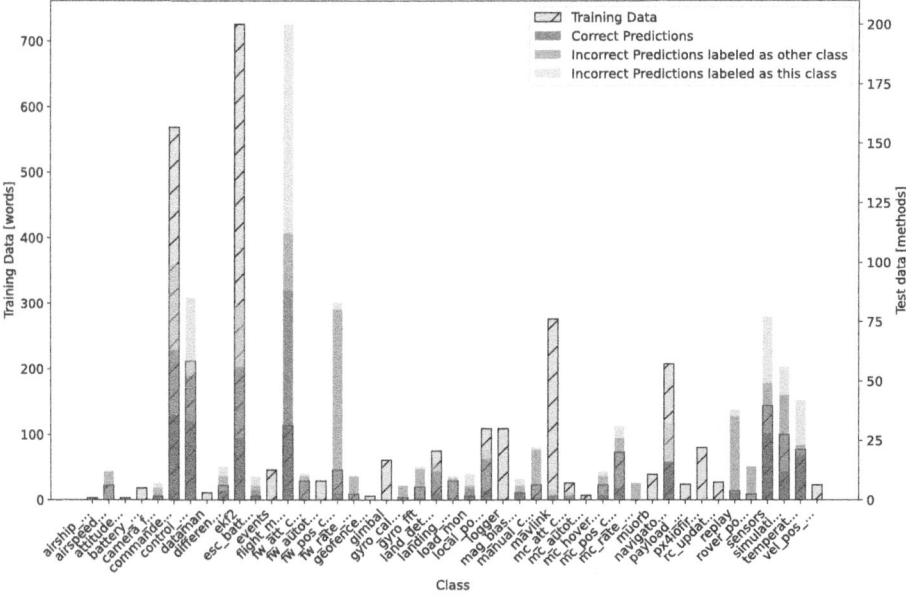

Fig. 5. Classification results in relation to the amount of training data per class.

Mapping Source Code to Software Architecture by Leveraging LLM's 145

not all configurations tested are entirely reasonable. As an example, in one configuration, we evaluated the word-embedding GloVe using a whole paragraph as a data point, which is not its intended use. Cases such as these were still included for completeness. The machine learning classifiers generally all performed comparably, except for SVM, which performed slightly worse for most configurations. However, this is likely a result of the hyperparameters used. The max_iter was set to only 25, chosen mainly because initial tests showed that SVM struggled to converge. If increased, SVM would likely perform as good as other classifers.

6.1 Threats to Validity

External Validity. The generated descriptions are an apparent external validity concern since they were generated using the code as input. However, the stepwise generation and paraphrasing have likely mitigated the risk of these being too similar to the code. That is, the language is too similar to the language of the code. If we would use a software system where corresponding documentation exists, it is possible that the documentation would still be very similar to the code. An initial review of different candidate systems showed that if such text existed, it often appeared to be a pseudo-code representation of the actual code. An issue that may have affected the result is that the method contents were not cleaned for potential references of the labels. It's possible that, for instance, methods contain the use of data structures defined in the class itself (written in the code by the format of $<classname>::<datastructurename>$). However, cleaning of class names from the user-defined identifiers data was performed, and those classifications performed at least equally to that of using method content. The results obtained should be regarded as specific to the dataset, i.e.,

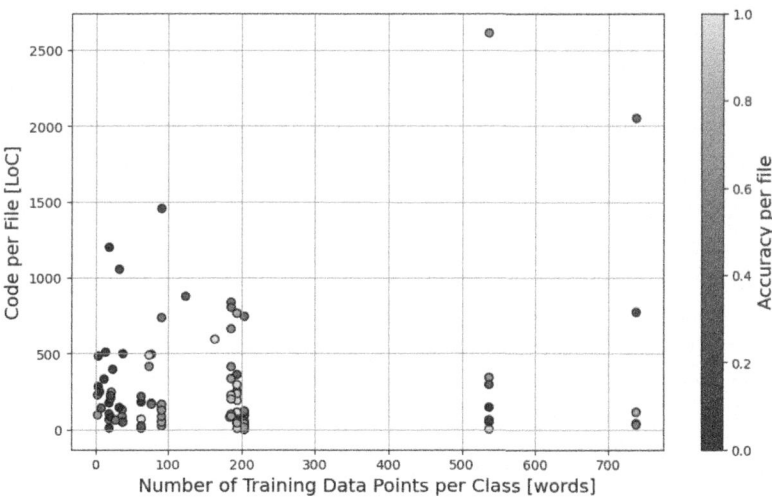

Fig. 6. Accuracy of predictions per file plotted in lines of code per file vs. training data per class.

the software system studied. Since the problem was approached as a language problem, the language of the code will affect the result. If the language of the code belonging to different modules is similar, it will be difficult to classify the code. Perhaps since the PX4 is an embedded system for a complete product, the terminology of the code is quite diverse, which is beneficial for classification.

Construct Validity. Using methods as classification tokens was selected since the resultant mappings would be valuable in a refactoring context. However, we have not evaluated how the classification of methods performs in this setting. Instead, we have attempted to classify the methods "back" to their original module. How the results compare to a realistic refactoring setting is not known. Evaluating for that would require using a target system where an old/legacy software edition is available, along with a new, refactored, software and accompanying documentation.

7 Conclusion and Future Work

With respect to **RQ1**, the results show that the LLMs outperform other common vectorization methods. Additionally, we conclude that augmenting the data has an apparent positive effect on accuracy. However, no particular classification method is outstanding. As for **RQ2**, from the analysis (see Sect. 6), we can see that having sufficient descriptions of modules in relation to source code is an important factor. Furthermore, we determine that the one classification method that does not need training, kNN, performs as well as any other classifier. This could be an effect of the difference between training and testing data, i.e., that the text of module descriptions is different from the source code. All combinations of vectorization methods with classification algorithms failed to find patterns during training that produced a result that outperformed kNN in testing. Thus, focusing on choosing and training a classifier may not be the effective way to attain higher accuracy.

The quality of the results is challenging to evaluate. If the proposed method is used in a software refactoring process, one would probably be satisfied with low accuracy provided that the precision is sufficiently high. One way to increase precision is to use classification models where one can choose not to classify a data point if the probability is below some threshold.

A factor that surely affected the accuracy of predictions is the high number of classes (45) and the imbalanced dataset. Each class definition(file) is described in around only 100 words, and some modules consist of many more class definitions than others. Applying up- and/or down-sampling might improve the accuracy of predictions. Involving a human to evaluate the data points that are difficult to predict is probably one sure way to increase performance. Such have previously been done, e.g., in [23], which involved a human evaluating a set of recommended seed mappings. A human can, e.g., aid the process after an initial prediction of methods by the ML model is done by manually labeling a set of points where the classifier struggles. We did not evaluate the use of single words from either training data (module descriptions) or testing data (code). The smallest sliding window of three words was chosen by the rationale that identifier names are often around three words long. Few identifier names consist of only a single word. Comparing each word in the descriptions to each word, e.g., all identifiers, and measuring similarity, could make the non-contextual vectorization perform differently.

This study has not used the structure of modules to any extent, purely the descriptions. However, using the relations of the target(intended) system may still be valuable to make better predictions. However, these properties may also be covered in the module descriptions, where one can include the relationship between modules.

Data Availability Statement. Data and Software developed to support the findings of this study are available at: https://github.com/nijoad/source-code-mapping-to-architecture/tree/main.

References

1. https://px4.io/
2. https://huggingface.co/blog/mteb
3. https://huggingface.co/sentence-transformers/all-MiniLM-L6-v2
4. https://huggingface.co/microsoft/unixcoder-base-nine
5. Abid, C., Alizadeh, V., Kessentini, M., do Nascimento Ferreira, T., Dig, D.: 30 years of software refactoring research: a systematic literature review. arXiv arXiv:2007.02194 (2020)
6. Somogyi, N., Kövesdán, G.: Software modernization using machine learning techniques. In: 2021 IEEE 19th World Symposium on Applied Machine Intelligence and Informatics (SAMI), pp. 000361–000365 (2021)
7. Alomar, E.A., Mkaouer, M.W., Newman, C.D., Ouni, A.: On preserving the behavior in software refactoring: a systematic mapping study. arXiv arXiv:2106.13900 (2021)
8. Alwosheel, A., van Cranenburgh, S., Chorus, C.G.: Is your dataset big enough? sample size requirements when using artificial neural networks for discrete choice analysis. J. Choice Model. **28**, 167–182 (2018)
9. Aniche, M.F., Maziero, E.G., Durelli, R.S., Durelli, V.H.S.: The effectiveness of supervised machine learning algorithms in predicting software refactoring. IEEE Trans. Software Eng. **48**, 1432–1450 (2020)
10. Baqais, A.A.B., Alshayeb, M.R.: Automatic software refactoring: a systematic literature review. Software Qual. J. **28**, 459–502 (2019)
11. Bittencourt, R.A., Santos, G.J.D., Guerrero, D.D.S., Murphy, G.C.: Improving automated mapping in reflexion models using information retrieval techniques. In: 2010 17th Working Conference on Reverse Engineering, pp. 163–172 (2010)
12. Christl, A., Koschke, R., Storey, M.A.: Automated clustering to support the reflexion method. Inf. Softw. Technol. **49**(3), 255–274 (2007). 12th Working Conference on Reverse Engineering
13. Cruciani, F., Moore, S., Nugent, C.: Comparing general purpose pre-trained word and sentence embeddings for requirements classification. In: 6th Workshop on Natural Language Processing for Requirements Engineering: REFSQ Co-Located Events 2023, vol. 3378. CEUR-WS (2023)

14. Devlin, J., Chang, M.W., Lee, K., Toutanova, K.: BERT: pre-training of deep bidirectional transformers for language understanding. arXiv:1810.04805 (2019)
15. Diaz-Pace, J.A., Berrios, R.C., Tommasel, A., Vazquez, H.C.: A metrics-based approach for assessing architecture-implementation mappings. In: Anais do XXV Congresso Ibero-Americano em Engenharia de Software, pp. 16–30. SBC, Porto Alegre, RS, Brasil (2022)
16. Dogra, V., et al.: A complete process of text classification system using state-of-the-art NLP models. Comput. Intell. Neurosci. **2022**, 1–19 (2022)
17. Florean, A., Jalal, L.: Mapping java source code to architectural concerns through machine learning. Master's thesis, Karlstad University (2021)
18. Florean, A., Jalal, L., Sinkala, Z.T., Herold, S.: A comparison of machine learning-based text classifiers for mapping source code to architectural modules. In: European Conference on Software Architecture (2021)
19. Guo, D., Lu, S., Duan, N., Wang, Y., Zhou, M., Yin, J.: Unixcoder: unified cross-modal pre-training for code representation. arXiv:2203.03850 (2022)
20. Hu, L., Liu, Z., Zhao, Z., Hou, L., Nie, L., Li, J.: A survey of knowledge enhanced pre-trained language models. arXiv:2211.05994 (2023)
21. Karakati, C.B., Thirumaaran, S.: Software code refactoring based on deep neural network-based fitness function. Concurrency Comput. Pract. Experience **35**(4), e7531 (2023)
22. Liang, M., Niu, T.: Research on text classification techniques based on improved TF-IDF algorithm and LSTM inputs. Procedia Comput. Sci. **208**, 460–470 (2022). 7th International Conference on Intelligent, Interactive Systems and Applications
23. Link, D., Behnamghader, P., Moazeni, R., Boehm, B.: Recover and relax: concern-oriented software architecture recovery for systems development and maintenance. In: Proceedings of the International Conference on Software and System Processes, ICSSP 2019, pp. 64–73. IEEE Press (2019)
24. Minaee, S., Kalchbrenner, N., Cambria, E., Nikzad, N., Chenaghlu, M., Gao, J.: Deep learning based text classification: a comprehensive review. arXiv:2004.03705 (2021)
25. Niu, C., Li, C., Luo, B., Ng, V.: Deep learning meets software engineering: a survey on pre-trained models of source code. arXiv:2205.11739 (2022)
26. Olsson, T., Ericsson, M., Wingkvist, A.: To automatically map source code entities to architectural modules with naive bayes. J. Syst. Softw. **183**, 111095 (2022)
27. Pace, J.A.D., Villavicencio, C., Schiaffino, S.N., Nicoletti, M., Vázquez, H.C.: Producing just enough documentation: an optimization approach applied to the software architecture domain. J. Data Semant. **5**(1), 37–53 (2016)
28. Pennington, J., Socher, R., Manning, C.D.: Glove: global vectors for word representation. In: Empirical Methods in Natural Language Processing (EMNLP), pp. 1532–1543 (2014)
29. PX4: Px4-autopilot/src/modules at main · px4/px4-autopilot. https://github.com/PX4/PX4-Autopilot/tree/main/src/modules
30. Savelka, J., Ashley, K.D.: The unreasonable effectiveness of large language models in zero-shot semantic annotation of legal texts. Front. Artif. Intell. **6**, 1279794 (2023)
31. Shah, K., Patel, H., Sanghvi, D., Shah, M.: A comparative analysis of logistic regression, random forest and KNN models for the text classification. Augmented Hum. Res. **5**(1), 1–16 (2020). https://doi.org/10.1007/s41133-020-00032-0
32. Sinkala, Z.T., Herold, S.: InMap: automated interactive code-to-architecture mapping. In: Proceedings of the 36th Annual ACM Symposium on Applied Computing,

pp. 1439-1442. SAC 2021, Association for Computing Machinery, New York, NY, USA (2021)
33. Sun, C., Qiu, X., Xu, Y., Huang, X.: How to fine-tune BERT for text classification? arXiv:1905.05583 (2020)
34. Wang, X., Wang, Y.: Sentence-level resampling for named entity recognition. In: North American Chapter of the Association for Computational Linguistics (2022)
35. Wang, Z., Pang, Y., Lin, Y.: Large language models are zero-shot text classifiers. arXiv:2312.01044 (2023)
36. Xie, Y., Lin, J., Dong, H., Zhang, L., Wu, Z.: Survey of code search based on deep learning. ACM Trans. Softw. Eng. Methodol. **33**(2), 1–42 (2023)
37. Yu, Y., et al.: Large language model as attributed training data generator: A tale of diversity and bias. arXiv:2306.15895 (2023)
38. Zhang, C., et al.: A survey of automatic source code summarization. Symmetry **14**, 471 (2022)

Mining for Sustainability in Cloud Architecture Among the Discussions of Software Practitioners: Building a Dataset

Sahar Ahmadisakha[✉][iD] and Vasilios Andrikopoulos[iD]

University of Groningen, Groningen, Netherlands
{s.ahmadisakha,v.andrikopoulos}@rug.nl

Abstract. The adoption of cloud computing is steadily increasing in designing and implementing software systems, thus it becomes imperative to consider the sustainability implications of these processes. While there has already been some academic research on this topic, there is a lack of perspective from practitioners. To bridge this gap, we utilize software repository mining techniques to examine 192 discussions among practitioners on the Software Engineering forum of the Stack-Exchange platform, aiming to build an annotated dataset containing cloud architectural discussions and to understand the current discussion on sustainability in cloud architecture. To identify these discussions, we first put together a list of terms indicating sustainability as the topic. Our initial findings indicate practitioners mainly focus on design aspects-analysis, synthesis, and implementation-while avoiding complex activities like evaluation and maintenance. Technical sustainability is emphasized, while the economic dimension has the most discussions exclusively focused on it. This contrasts with previous academic literature, which highlighted environmental sustainability.

Keywords: cloud computing · sustainability dimensions · mining software repository · software architecture

1 Introduction

Although sustainability in software engineering has a longstanding history, its integration into software architecture is a relatively recent development [17]. This shift has been profoundly influenced by the rise of cloud computing, which refers to the provision of internet-accessible computing resources [19]. Notably, this emergence has foregrounded the environmental sustainability aspect, primarily due to the substantial energy consumption of data centers, which serve as the primary hosts for cloud infrastructures [3].

This work is partly funded by the project SustainableCloud (project number OCENW.M20.243) of the research program Open Competition Domain Science by the Dutch Research Council (NWO).

© The Author(s), under exclusive license to Springer Nature Switzerland AG 2024
A. Ampatzoglou et al. (Eds.): ECSA 2024, LNCS 14937, pp. 150–166, 2024.
https://doi.org/10.1007/978-3-031-71246-3_14

Extensive research endeavors have already been undertaken at the intersection of software architecture and sustainability, aiming to address sustainability [16,28,29] or evaluate it [15] and understand it from the practitioners' perspective [8,9]. However, there have been limited efforts to explore the implications of cloud computing on this intersection like [1] investigating all dimensions from an academic view and [30] focusing on environmental dimension from the practitioners' perspective.

The **objective** of this study is therefore to ascertain the perspectives of software practitioners regarding sustainability dimensions within cloud architectural discussions. The dimensions include *technical, economic, environmental,* and *social,* concerning *software longevity, capital preservation, natural resources conservation,* and *community continuity,* respectively [17]. We aim to identify how sustainability is recognized within these discussions and to what extent its various dimensions are addressed. To achieve our objective, we utilize software repository mining techniques to analyze discussions among practitioners on the *Software Engineering (SE)* forum of StackExchange, and build a dataset that contains cloud architectural discussions along with their associated sustainability dimensions. Despite the presence of previous studies [7,12,25] that explore architectural knowledge in *Stack Overflow*, we specifically focus on the SE forum and do not include other forums such as *Stack Overflow* or *Code Review*, as they yielded a small useful sample in our pilot query. We believe that the selected forum is appropriate enough since SE is dedicated to software engineering and it is more likely to contain discussions on architectural topics.

The primary **contributions** of this study are therefore twofold: *Firstly,* the establishment of an *initial set of terms* for conducting searches related to sustainability in corpora. This set could be used for both systematic surveys of the literature, and for mining studies in the future. *Secondly,* the creation of a validated and annotated *dataset of practitioners' discussions* related to the various sustainability dimensions in conjunction with cloud architecting, identifying specific quality requirements and architecting phases in them. This annotated dataset provides a valuable resource for further academic research on the topic. It also offers insights to researchers and practitioners, highlighting the current state and elicitation methods of sustainability dimensions in architectural discussions.

The paper continues with an expanded study design in Sect. 2, followed by results in Sect. 3, and discussion in Sect. 4. Threats to validity are addressed in Sect. 5, with related works in Sect. 6. Conclusions and future work are presented in Sect. 7, with a statement on data availability in Sect. 8.

2 Study Design

As the goal of this work is to establish the practitioners' perspective on the topic of sustainability in architecting cloud-based software systems, we opt to utilize the established checklist provided by the ACM SIGSOFT Empirical Standards for Software Engineering [22] as the methodological backbone to conduct a mining software repositories (MSR) study, as outlined in the following.

2.1 Research Aims and Questions

The objective of this study using the Goal-Question-Metric [27] formulation is: *Analyze* the experiences and opinions of software practitioners in discussion forums *for the purpose of* understanding and building a dataset from the practitioners' perspective about sustainability, *with respect to* the associated dimensions *from the point of view* of the software practitioners *in the context of* architecting cloud-based software systems.

This investigation is conducted through the lens of **quality requirements** by examining discussions in software repositories that we call "data points". The stated goal leads to two research questions:

RQ1: *Which architectural phases are considered in the discussions?*

Since SE is not a forum specific to software architecture and is generally about software engineering, and in order to ensure a comprehensive analysis based on architectural discussions, it is crucial to identify only the relevant discussions. For this purpose, we use data points that encompass *at least one phase of the architecting life cycle* so-called *analysis, synthesis, evaluation, implementation, and maintenance* [24], to be able to elicit the necessary architectural knowledge such as quality requirements. By means of identifying these phases, we also aim to understand the evolution of cloud architecture discussions over time.

RQ2: *What is the state of the discussions with regard to the sustainability dimensions?*

At the same time, we also aim to comprehend the status of sustainability considerations within software repositories through discussions among software practitioners. This endeavor is essential for understanding the communication of sustainability aspects within these repositories and enables us to compare those findings against those previously extracted from the literature in existing work [1].

2.2 Dataset Building

This section of the study design encompasses three steps that we further elaborate on, namely: *Repository selection*, *Query building*, and *Data point selection*. The overall process is summarized later in the paper in Fig. 2.

Step 1: Repository Selection. For our study's objectives, we already selected Q&A platforms, notably Stack Exchange (SEx), as it offer discussions on software engineering topics. Among the SEx forums, we focused on the SE forum. We chose this because SEx lacks a specific platform dedicated to discussing issues or ideas surrounding software systems architecture and because it is a forum for software engineering discussions. Moreover, recent studies [11] have indicated that Q&A websites, such as Stack Exchange (and its forums), are admitted by practitioners as the most valuable source of architectural information.

Table 1. Below are the terms included in our mining query. We removed the "*" sign from certain terms like `architect*` or `efficien*`, as it is handled by "%" in the actual query. Please note that only a portion of the sustainability terms are included.

Category	Query Terms				
Architecture	architect	design	tactic	pattern	best practice
	model	principle	analysis	synthesis	maintenance
	evaluation	implementation			
Cloud	AWS	EC2	S3	Aurora	Azure
	Azure DevOps	Azure Virtual	Azure Blob	Google Cloud	GCP
	Google Compute	GCE	Google App	GAE	GCS
Sustainability	sustain	technic	econom	soci	environment
	cost	effort	efficien	evol	resource
	performance	continu	complexity	consum	...

Query Terms Identification. To collect data from the above-mentioned repository, we utilized a query through the SEx data explorer (SEDE)[1], same as [23]. Formulating this query involves selecting appropriate terms tailored to our study. Due to the goal of this work, our query revolves around three key concepts: *software architecture*, *cloud computing*, and *sustainability*. We incorporated multiple terms for each concept to maximize the inclusion of relevant data points. In the following, we discuss the rationale behind the selection of the query terms as they appear summarized in Table 1. The full list of terms and the query itself are available in the replication package of our study (see Sect. 8).

Query Terms Related to Software Architecture: For this we adopt the `architect*` term and choose the main software architecture's terms from [6]. We also add the architecting phases defined in [24] as other terms to this part of the query. For more information about the query terms see Table 1.

Query Terms Related to Cloud Computing: We include the term `cloud` and the names of the top three cloud providers based on Flexera's State of the Cloud Report, 2024: Amazon Web Services, Microsoft Azure, Google Cloud Platform, along with their abbreviations (AWS, Azure, GCP), and their three most commonly used cloud services in the query. Sources for top services for AWS, Azure, and GCP are in the hidden links.

Query Terms Related to Sustainability: The most challenging aspect of constructing the search query lies in selecting the appropriate set of query terms for sustainability. To our knowledge, there is no predefined list of keywords specifically tailored for searching sustainability topics that cover all its dimensions. Moreover, our pilot query with the term `sustain*`, similar to most secondary studies, yielded zero results. This suggests the need for additional search terms

---

[1] https://data.stackexchange.com/ [Accessed: 18 March 2024].

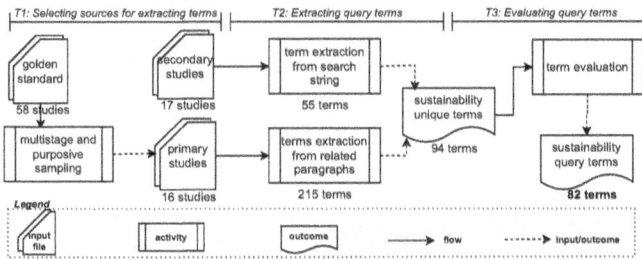

Fig. 1. The process of extracting sustainability-related terms for further query building step. **82** such terms are extracted.

for sustainability. Existing studies [13,15] often rely on terms like "green", "environment*", or "sustainab*", but previous research highlights limitations in capturing sustainability content solely through these terms [1]. This challenge stands as a significant burden in sustainability research. This led us to explore alternative terms for investigating sustainability topics, inspired by previous work in mining software repositories from an energy efficiency aspect [2]. To this effect, we initiated a three-step process (T1 to T3) to identify additional terms beyond basic sustainability terms, as outlined in Fig. 1; more specifically:

T1. Sources for Extracting Sustainability Query Terms
We initiate our term pool by identifying related terms from two sources: primary, and secondary literature studies. A list of 17 *secondary studies* (see replication package in Sect. 8 for the full list) was chosen based on their coverage of sustainability in software engineering and architecture research, and on whether they were conducted in a systematic manner. For *primary studies*, we use the dataset of systematic mapping study of Andrikopoulos et al. [4] in two ways: as an additional source of terms come from primary studies, and as a *gold standard* for assessing whether the terms we selected are suitable for retrieving sustainability-related works (T3, below).

We start by sampling the set of primary studies in [4]. To sample the dataset, we cluster studies based on their addressed sustainability dimensions. We employ various techniques, including *multistage* sampling guided by methodologies outlined in [5], and *purposive* sampling, integrating both probability and non-probability sampling methods. Given the variability in the number of sustainability dimensions addressed by primary studies (ranging from 1 to 4), we prioritize studies covering a single dimension. This allows us to select 3 primary studies per dimension, as we believe they offer more tailored and specific terms. Down this road, we faced challenges with the economic and social dimensions. No primary studies were exclusively dedicated to the economic dimension in [4]. To address this, multistage sampling was used. We selected a *cluster* including studies addressing 2 dimensions, one of which relates to the economic dimension, and chose 3 primary studies *randomly* from them in the cluster. Regarding the social dimension, only 1 study focused solely on it, which we included in our selection. We also identified 3 studies addressing the social dimension along

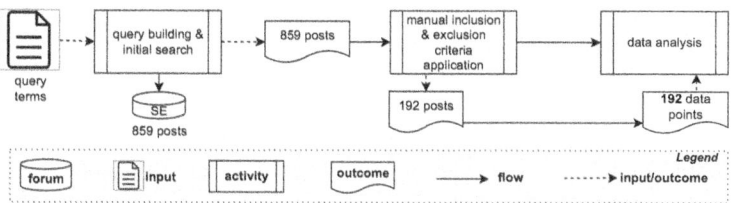

Fig. 2. The process of extracting and filtering data points from SEDE to analysis. **859** data points are extracted among which **192** data points are accepted.

with another dimension and *randomly* selected 2 of them. For the technical and environmental dimensions, we *randomly* chose 3 primary studies for each among the studies that address solely the technical or environmental dimension.

Finally, we bolstered our initial selection of 12 primary studies by adding 4 more studies through *purposive sampling*. These additional studies covered all four sustainability dimensions. We included them because comprehensive coverage of all dimensions enhances the focus on sustainability, increasing the likelihood of identifying relevant terms.

T2. Extracting Sustainability Query Terms
Upon completing the process of identifying relevant studies for extracting sustain-ability-related terms, we proceed to extract query terms from these sources. In this process, before doing any extraction, we initially select the terms `sustain*`, `technic*`, `econom*`, `environment*`, and `soci*` as **basic terms** of sustainability and its dimensions. From the 17 secondary studies, we extract sustainability-related terms from their search strings. For the 16 selected primary studies, we thoroughly surveyed the text to identify paragraphs discussing **basic terms** and gathered any technical, economic, social, or environmental terms mentioned. We also examine the text for sustainability-related concepts, such as "responsible resource use for green computing," aligning with the environmental dimension. During term extraction, we highlighted relevant lines from papers, ensuring each term was only extracted once per document. We then normalized the terms, reducing 270 initial terms to **94 unique terms**, and mapped both initial and normalized terms to sustainability dimensions, as detailed in the replication package.

T3. Evaluating Sustainability Query Terms
As a way of evaluating the resulting set of terms, we check whether we can retrieve with them all 58 primary studies from our gold standard dataset. We search all 94 terms across these studies (excluding their references sections). By using all 94 terms, we indeed can retrieve all 58 primary studies. However, we could see that while some terms like `time`, `sustain*`, `change`, `user`, `quality`, `maint*`, and `value` seem promising in retrieving 100–88 percent of the studies, except for `sustain*`, they lack specificity and may vary in meaning depending on context. Therefore, we opt to eliminate them and search for

Table 2. Inclusion and exclusion criteria. I*N* stands for inclusion criterion number and E*N* stands for exclusion criterion number.

ID	Criteria	Explanation
I1	Zero or positive Score data point	We want to include data point that has not been given negative scores from other practitioners
I2	Data point is about an architectural discussion	We want to keep data points that necessarily have an architectural discussion
E1	False positive data point	We have to reject the data point if it includes no data or perception on cloud computing or is out of the topic
E2	PostTypeId > 2	In Posts tables, we want to keep data points that are either question or answer. (PostTypeId = 1 or 2 respectively)

a *minimal set* within the remaining terms. A minimal set comprises the fewest terms needed to retrieve all primary studies. To find one, we sort terms by their coverage of primary studies in descending order of appearance frequency and sequentially add these terms to the minimal set until all studies from [4] can be retrieved using the set. A minimal set of **82 terms**, available in the replication package, is constructed through this process.

Step 2: Query Building. After identifying the appropriate terms for the query, we build the query as follows:

([software architecture-related terms in OR]) AND ([cloud computing- related terms in OR]) AND ([sustainability-related terms in OR])

using the terms in Table 1 for software architecture and cloud computing, and the 82 terms from T3 in the previous step for sustainability, as shown in Fig. 2.

Step 3: Data Point Selection. After executing the query we proceed by selecting the necessary columns from the Posts table in the SEDE results. We retain 5 out of the 23 available columns, namely: Id which is a unique identifier for each post; PostTypeId which is the type of post (e.g., question, answer); CreationDate which is date and time when the post was created; Score which is the total score of the post, contains the net sum of upvotes and downvotes; and Body which is the main content of the post, including text, code snippets, and other relevant information. Thereafter, we apply our inclusion and exclusion criteria, outlined in Table 2, to filter relevant data points. Any post meeting all inclusion criteria and none of the exclusion criteria progresses to the next stage which is analysis. After applying our criteria to the **initial 859 data points**, we focus on the remaining **192 accepted data points** to address our research questions. To note: we accept only 22% of posts. Most rejections (64%) are due to false positives (E1). Only 10% are not architecturally relevant (I2), and the remaining 4% involve posts with negative scores or inappropriate IDs (I1, E2). With only 10% of discussions being architecturally irrelevant, the Software Engineering forum on SEx *can be* ideal for architectural discussions.

2.3 Data Extraction and Initial Analysis

We *manually* extract and organize the data in a spreadsheet, mainly focusing on the main discussions in the posts' Body column.

RQ1. To address this research question, we read each data point to determine which architecting phase it pertains to based on the activities reported in [24]. This evaluation (and the next one for RQ2) is primarily conducted manually by one researcher, with random checks by a second researcher to ensure consistency (two researchers involved in this process).

RQ2. To address our research question, we identify the sustainability dimensions discussed within each data point. Since data points do not directly address sustainability dimensions, we first extract the quality requirements (QR) mentioned in the entire post. We then map these QRs to sustainability dimensions using results in [1] and the sustainability quality model outlined in [8] as a guide (the mapping table is available in the replication package). However, 72 data points do not explicitly mention any QRs. For these, we assign appropriate QRs by exercising our judgment (indicated by coded-qr in the dataset, together with the post's relevant part). As an illustration, consider a part of data point ID 122098 which implicitly talks about fault tolerance and scalability:

> ...*To me the key point is the cloud is a set of cheap, unreliable resources.* ***Your solution has to be built to keep running when the resources fail, and have a scalable architecture that can utilize additional resources....***

As an illustration of how QR to sustainability dimension mapping happens, consider data point ID 432796:

> ..*This doesn't feel quite right, am I missing a **security**/privacy measure here? I'm thinking about pushing this app soon, potentially marketing it but, at first, simply handing the APK to family and friends for wider use and testing of the beta - I'll have access to their most personal memories. Is it a matter of **trust** or is there some way I can implement a certificate or something somehow for?..*

This example demonstrates the presence of two quality requirements: security and trust. As per the mappings in [1,8], security contributes to the technical and social dimensions, while trust contributes to the technical, economic, and social dimensions. Consequently, this data point reflects aspects of all dimensions except for the environmental one. Moreover, it indicates that the discussion primarily focuses on technical and social aspects, as evident from the quotation.

A part of data point ID 360022 as another example of this mapping is presented:

...First Decision: Self Hosted vs the Cloud: The cloud allows you to use AWS S3 or whatever equivalent blob storage for your cloud provider. ***This solution only charges you for the storage you use, and cloud blob storage provides both the scale and performance needed to scale as your application grows...***

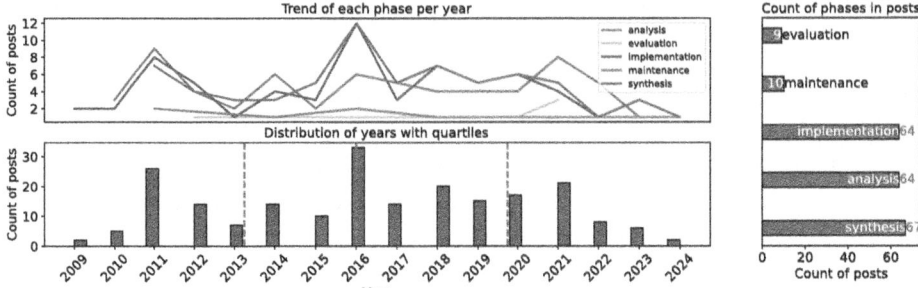

Fig. 3. From top-left–*Upper plot*: architecting phases addressed along the years. *Lower plot*: Cloud architecture discussions across years. *Vertical plot*: Architecting phases counts that are addressed in the SE forum posts discussions

This quotation discusses a design decision for selecting the hosting infrastructure for an application, highlighting the cloud's ability to scale with growth and charge only for used resources. This underscores the economic importance of `scalability` in resource utilization. The economic contribution of scalability is also noted by [8], and we incorporated this in the mapping.

3 Preliminary Results and Findings

In Fig. 3, we illustrate the distribution of discussions over 16 years. While the lower plot suggests a binomial distribution, the data from the first half occurred within nearly eight years, centered around 2016, which marks the second quartile (Q2) of the data. This implies a slightly higher frequency of architectural discussions during the early years of cloud computing emergence.

3.1 RQ1: Architectural Discussion Coverage

Based on the architectural phases identified during the data extraction process, one observation pertains to the distribution of discussions across architectural phases emerges, as illustrated in Fig. 3. Notably, there appears to be a balanced representation of the analysis, synthesis, and implementation phases. However, the relatively limited presence of maintenance and evaluation discussions may arise from various factors. Regarding maintenance, it might be due to the comprehensive support provided by cloud service providers. In the context of cloud computing, tasks such as data center management, operating system updates, security patching, load balancing, resource monitoring, data backup, and disaster recovery are typically handled by the cloud infrastructure, potentially reducing the need for extensive discussion on maintenance activities.

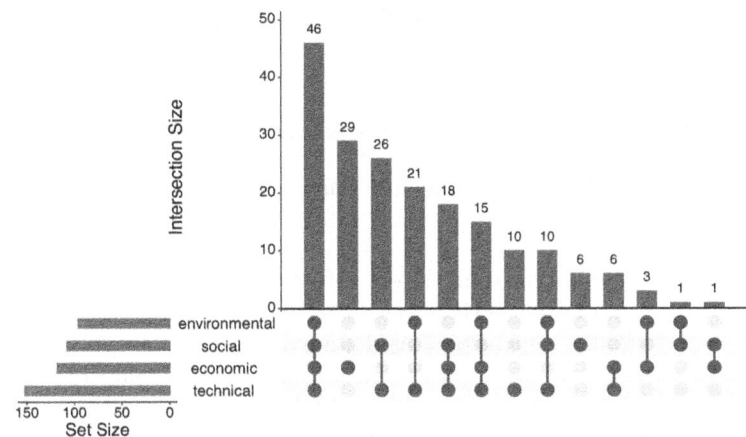

Fig. 4. Discussed sustainability dimensions along with their intersections.

In terms of the evaluation phase, it appears that assessing cloud architectures may not be the primary focus of interest. Rather, there is a greater emphasis on discussing how to design and deploy architectures for cloud environments. This suggests that practitioners might be primarily engaged in discussing the appropriate way for building and migrating applications to operate effectively within cloud environments. Consequently, having frequent discussions about analysis, synthesis, and implementation phases *may* be attributed to this ongoing exploration and advancement of cloud-based architecture practices.

Another point can be made by examining the architectural phases over the years from Fig. 3. The upper plot in the figure indicates similar behaviors between the analysis and implementation phases before Q2 (which is around 2016), with a subsequent divergence after Q2 where the implementation phase continues with synthesis. We speculate that the fact that implementation phases have evolved with analysis somehow near 2016 and then continued with synthesis might be related to the first official demonstration of AWS's well-architected framework that emerges at 2016, as supported by the biggest cloud provider. This framework may have given architects more means to do implementation and build on top of the cloud providers.

3.2 RQ2: State of the Sustainability Discussions

By extracting quality requirements from the discussions and mapping them to sustainability dimensions, it becomes apparent that a discussion may encompass

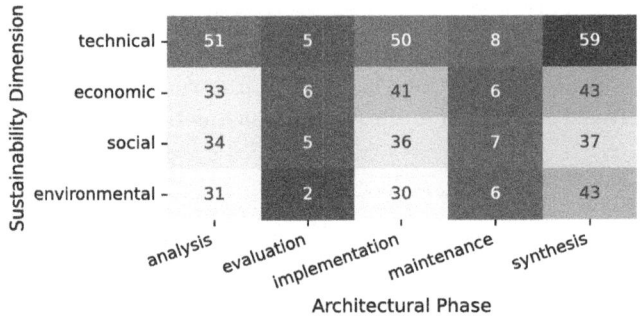

Fig. 5. The co-occurrence of sustainability dimensions with architectural phases.

one or multiple sustainability dimensions. After doing the mapping for all data points, the resulting analysis reveals the distribution of data points containing any of the sustainability dimensions, as depicted in Fig. 4. Notably, the technical dimension emerges as the most commonly mapped, which aligns with the tendency of practitioners to mainly address and tackle technical aspects.

Figure 4 also illustrates the intersections among various dimensions that we call combinations. Notably, these intersections closely correspond to those revealed in academic literature in [1] except for the combination of social, economic, and environmental dimensions (soc-eco-env), and solo environmental (env) which are not detected in the practitioners' discussions. In the current study, we identify four additional combinations (env-soc, env-soc-tec, tec-soc-eco-env), and (soc) which are not detected in [1]. We share 11 combinations with that reported in [1]. Among the newly found intersections, one notable example is the combination of technical, social, economic, and environmental dimensions (tec-soc-eco-env), which surprisingly emerges as the most frequent intersection. This is one of the combinations that is not found at all in [1]. Upon closer examination of the figure, we observe that after the tec-soc-eco-env combination, the economic dimension emerges as the second most popular. Next frequent combinations predominantly involve the technical-social (tec-soc) and technical-environmental (tec-env) dimensions. This observation suggests a potential correlation between the technical dimension and both social and environmental dimensions.

Takeaway 1. *The economic dimension is primarily addressed individually, indicating that it not only accounts for the highest number of data points solely focused on this dimension, but also makes a substantial contribution to the majority of the discussions. This is evident from its position as the second dimension in terms of set size in Figure 4. This underscores the significance of economic considerations in the discussions among practitioners.*

Based on the data we have collected thus far from the dataset and initial analysis, the final aspect we wish to present concerns the intersection of architectural phases and sustainability dimensions, as illustrated in Fig. 5. The figure indicates that all sustainability dimensions are addressed across all architectural phases; however, the frequency of their co-occurrence in the evaluation and maintenance phases is notably low, for all dimensions. As is high for analysis, synthesis, and implementation across all dimensions. In summary, sustainability dimensions do not appear to significantly impact the extent to which each architectural phase is communicated. This observation may be attributed to the specific forum selected or the context of cloud computing.

4 Discussion

Based on our findings, we would like to identify some points of discussion that require further consideration.

To start off, a list of terms for searching about sustainability is presented in this study. This list corresponds to the shortage of suitable terms for researching sustainability, as highlighted by [1]. *Our approach, which entails identifying sustainability-related query terms, does not claim to cover every potential study or capture all sustainability-related terms. Instead, it represents an initial effort to identify terms for further sustainability research which is sufficient to fulfill our needs in this study.* Further efforts to both augment the list and further examine it are part of our future work. The question, however, arises: if the list is adequate for sustainability research, why are most data points rejected? The answer lies in the inclusion and exclusion criteria we have set for this study. We do not evaluate data points based on their direct connection to sustainability, as this is achieved through QR mapping. The posts, as discussed and presented in Sect. 2, are mostly rejected because they do not provide a cloud architectural discussion. These rejections stem from various issues, such as inappropriate usage of cloud terms, the inclusion of cloud terms just within links, mentioning only a cloud service without substantive discussion, and providing definitions of cloud computing, etc. See this sample that `cloud` term is used irrelevantly:

> *Time to step down from the **clouds**. Composites actually describe the meat of the code. I'm talking about classes, methods, objects, functions, prototypes...*

The findings pertaining to the most addressed dimension in Sect. 3.2 diverge from those reported in the academic literature [1] in the same context of cloud architecting. Previous research highlights the notable significance of the environmental dimension in academia, with the technical dimension ranking third. This divergence could be attributed to the specific repository under analysis or may underscore differences in how sustainability is perceived and prioritized by practitioners compared to academics. It is important to note that the results presented here are comparable only to those of [1], as it is the sole study sharing a similar context with the current investigation, specifically in terms of analyzing *cloud architectural* discussions. We would like to mention one of the takeaways:

Takeaway 2. *In the context of cloud computing, academia tends to prioritize environmental sustainability, while practitioners exhibit a stronger focus on technical sustainability. This indicates potential variations in the prioritization* and communication of sustainability dimensions between academic discourse and practical implementation within cloud architecting.

Another interesting point to add to the discussion is that in 2021, the AWS well-architected framework (that we first mentioned in Sect. 3.1) integrated a sustainability pillar, with a strong emphasis on the environmental dimension of cloud-based software. Almost all of the major cloud providers, including Microsoft Azure and Google Cloud Platform, have developed similar frameworks. However, we do not observe any increase in the number of discussions related to sustainability in the years following 2021, and definitely no increase in the environmental sustainability dimensions. *This may indicate a potential disinterest by the practitioners in the topic of environmental sustainability.* Future work to clarify this disparity is required.

Finally, we believe that our approach of extracting architectural discussions in general—that is, without considering sustainability or cloud computing as part of the picture—from the SEx platform has been effective, with a minimal rejection rate of 10%. *However, considering the comprehensive architecture centers and sustainability best practices offered by major cloud providers like AWS and Azure, leveraging their community forums (e.g. AWS, Azure) instead of the general-purpose SE forum at SEx may offer greater advantages.* The challenge lies in the lack of straightforward methods to obtain forum data through downloads or queries from these forums, prompting us to continue mining data from the SEx for the time being.

5 Threats to Validity

External Validity: Utilizing only one forum of SEx, meaning SE, creates a potential external validity threat. However, the most relevant forum in this Q&A platform to our study is SE, and there are recent studies in our field like [23] that similarly take SE as one of three relevant forums of SEx. We excluded Stack Overflow (SO) and Code Reviews (CodeR) due to their emphasis on programming issues, which are not in our interest. Despite expecting higher results from SO and similar results from CodeR in comparison with the SE forum (considering their whole number of questions and answers to be higher and equal with SE), our tailored query yielded fewer data entries from them (SO and CodeR), indicating their potential irrelevancy to our topic. Access to the Stack Exchange Data Explorer (SEDE) was rechecked on March 18th, 2024, to update the study's results. For reproducibility, the complete replication package of this research is made available online, as discussed in Sect. 8. However, a potential concern arises regarding the data acquisition method, since similar to prior recent studies like [23] we query the SEDE which may not retrieve all available data due to caching and request capping. To address this concern, our query was executed multiple times using different browsers, users, and timestamps to minimize potential data loss.

Internal Validity: We employ a predefined set of terms for querying data points, a method proven effective in energy-related software mining (e.g., [10,

21]). To mitigate potential drawbacks, such as false positives, *we manually analyzed all data points to prevent any false positive inclusions*. To address potential false negatives from the initial query, we utilized over 90 different queries, systematically adding each sustainability-related term. The number of query results becomes constant around the term `flexibility`, indicating saturation. However, to ensure comprehensive coverage, we included the remaining terms in the final query. Details regarding the saturation point and results per query are provided in the replication package. While we have just created the initial query terms for sustainability, we cannot guarantee that we have eliminated all possible false negatives. However, by using various queries and employing terms from relevant literature, we aimed to minimize this risk.

6 Related Works

In the realm of mining software repositories, extensive research has explored various aspects of software development, yet a notable gap persists: the investigation of sustainability dimensions. While numerous studies have delved into energy-related topics in robotics and Android applications, as well as broader aspects of the software development life cycle, sustainability dimensions remain relatively unexplored. For example, Cruz et al. [10] examined the impact of energy-oriented code changes on the maintainability of Android applications, providing insights into the intersection of energy efficiency and application development. Similarly, Albonico et al. [2] and Malavolta et al. [18] have recently investigated energy-related practices within robotics software in related repositories including StackOverflow, addressing gaps in knowledge and offering comprehensive analyses of energy efficiency in robot operating systems.

In a different track, efforts by Pinto et al. [21] on StackOverflow and Moura et al. [20] explored software energy consumption concerns among application programmers, providing valuable insights into energy-aware strategies for improving energy efficiency in real-world applications. A bit on social aspects and architectural concepts, in parallel, the study by Tizard et al. [26] discusses software project failures attributed to a lack of user feedback and missed requirements, proposing mining product forums as a solution to address this challenge. Koetter et al. [14] delve into the intersection of product quality requirements and social aspects within software projects. They analyze commit history to extract success and failure factors, providing recommendations to enhance project quality, including maintainability. While these studies touch upon architectural, social, or energy-related aspects that may intersect with sustainability, none explicitly address sustainability dimensions in the context of software repositories, particularly within cloud architectural discussions. Recognizing this gap, we embarked on the journey *to construct a dataset* focused explicitly on sustainability dimensions, laying the groundwork for future works in this crucial area of software architecting.

7 Conclusion and Future Work

We analyzed data points from the SEx repository (specifically from the SE forum) *to understand and build a dataset from practitioners' perspectives on sustainability in cloud architectural discussions*, considering the growing importance of sustainability in such systems. Initially, we extracted architectural phases and quality requirements from each data point (post) and then annotated them with the respective sustainability dimension(s) by means of quality requirements mapping to sustainability dimensions. We build a dataset consisting of posts containing cloud architectural discussions, together with the extracted QRs and associated sustainability dimensions. Our initial findings suggest that the SE forum holds promise for architectural discussions and that there is a notable emphasis on technical and economic sustainability. Furthermore, practitioners engaged in cloud architecting are still in the process of discussing how to design, implement, and deploy applications to the cloud. Although we found a suitable forum for architectural discussions (SE), we recognize the need for additional sites and repositories such as those discussed in Sect. 4. Additionally, we propose revisiting the sustainability annotation process based on quality requirement mappings through consensus building to enhance the robustness of our findings.

8 Data Availability

Data associated with this research is publicly available online in the replication package provided at the link.

References

1. Ahmadisakha, S., Andrikopoulos, V.: Architecting for sustainability of and in the cloud: a systematic literature review. Inf. Softw. Technol. **171**, 107459 (2024)
2. Albonico, M., Malavolta, I., Pinto, G., et al.: Mining energy-related practices in robotics software. In: 2021 IEEE/ACM 18th International Conference on Mining Software Repositories (MSR), pp. 483–494. IEEE (2021)
3. Andrikopoulos, V., Lago, P.: Software sustainability in the age of everything as a service. In: Aiello, M., Bouguettaya, A., Tamburri, D.A., van den Heuvel, W.-J. (eds.) Next-Gen Digital Services. A Retrospective and Roadmap for Service Computing of the Future. LNCS, vol. 12521, pp. 35–47. Springer, Cham (2021). https://doi.org/10.1007/978-3-030-73203-5_3
4. Andrikopoulos, V., et al.: Sustainability in software architecture: a systematic mapping study. In: 2022 48th Euromicro Conference on Software Engineering and Advanced Applications (SEAA), pp. 426–433. IEEE (2022)
5. Baltes, S., Ralph, P.: Sampling in software engineering research: a critical review and guidelines. Empir. Softw. Eng. **27**(4), 94 (2022)
6. Bass, L., Clements, P., Kazman, R.: Software Architecture in Practice, 4th Edition. SEI series in software engineering, Addison-Wesley Professional, Boston (2021)
7. Bi, T., et al.: Mining architecture tactics and quality attributes knowledge in stack overflow. J. Syst. Softw. **180**, 111005 (2021)

8. Condori-Fernandez, N., Lago, P.: Characterizing the contribution of quality requirements to software sustainability. J. Syst. Softw. **137**, 289–305 (2018)
9. Condori-Fernandez, N., Lago, P., Luaces, M.R., Places, Á.S.: An action research for improving the sustainability assessment framework instruments. Sustainability **12**(4), 1682 (2020)
10. Cruz, L., Abreu, R., Grundy, J., et al.: Do energy-oriented changes hinder maintainability? In: 2019 IEEE International conference on software maintenance and evolution (ICSME), pp. 29–40. IEEE (2019)
11. de Dieu, M.J., Liang, P., Shahin, M.: How do developers search for architectural information? an industrial survey. In: 2022 IEEE 19th International Conference on Software Architecture (ICSA), pp. 58–68. IEEE (2022)
12. de Dieu, M.J., et al.: Characterizing architecture related posts and their usefulness in stack overflow. J. Syst. Softw. **198**, 111608 (2023)
13. García-Mireles, G.A., Moraga, M.Á., García, F., et al.: Interactions between environmental sustainability goals and software product quality: a mapping study. Inf. Softw. Technol. **95**, 108–129 (2018)
14. Koetter, F., et al.: Assessing software quality of agile student projects by data-mining software repositories. In: CSEDU (2), pp. 244–251 (2019)
15. Koziolek, H.: Sustainability evaluation of software architectures: a systematic review. In: Proceedings of the joint ACM SIGSOFT conference–QoSA and ACM SIGSOFT symposium–ISARCS on Quality of software architectures–QoSA and architecting critical systems–ISARCS, pp. 3–12 (2011)
16. Lago, P.: Architecture design decision maps for software sustainability. In: 2019 IEEE/ACM 41st International Conference on Software Engineering: Software Engineering in Society (ICSE-SEIS), pp. 61–64. IEEE (2019)
17. Lago, P., Koçak, S.A., Crnkovic, I., Penzenstadler, B.: Framing sustainability as a property of software quality. Commun. ACM **58**(10), 70–78 (2015)
18. Malavolta, I., Chinnappan, K., Swanborn, S., et al.: Mining the ROS ecosystem for green architectural tactics in robotics and an empirical evaluation. In: 2021 IEEE/ACM 18th International Conference on Mining Software Repositories (MSR), pp. 300–311. IEEE (2021)
19. Mell, P., Grance, T., et al.: The nist definition of cloud computing (2011)
20. Moura, I., et al.: Mining energy-aware commits. In: 2015 IEEE/ACM 12th Working Conference on Mining Software Repositories, pp. 56–67. IEEE (2015)
21. Pinto, G., Castor, F., Liu, Y.D.: Mining questions about software energy consumption. In: Proceedings of the 11th Working Conference on Mining Software Repositories, pp. 22–31 (2014)
22. Ralph, P., Ali, N.b., Baltes, S., et al.: Empirical standards for software engineering research. arXiv preprint arXiv:2010.03525 (2020)
23. Tahir, A., Dietrich, J., Counsell, S., et al.: A large scale study on how developers discuss code smells and anti-pattern in stack exchange sites. Inf. Softw. Technol. **125**, 106333 (2020)
24. Tang, A., et al.: A comparative study of architecture knowledge management tools. J. Syst. Softw. **83**(3), 352–370 (2010)
25. Tian, F., Liang, P., Babar, M.A.: How developers discuss architecture smells? an exploratory study on stack overflow. In: 2019 IEEE International Conference on Software Architecture (ICSA), pp. 91–100. IEEE (2019)
26. Tizard, J.: Requirement mining in software product forums. In: 2019 IEEE 27th international requirements engineering conference (RE), pp. 428–433. IEEE (2019)
27. Van Solingen, R., Basili, V., Caldiera, G., Rombach, H.D.: Goal question metric (GQM) approach. Encyclopedia of software engineering (2002)

28. Venters, C.C., et al.: Software sustainability: research and practice from a software architecture viewpoint **138**, 174–188 (2018)
29. Venters, C.C., et al.: Sustainable software engineering: reflections on advances in research and practice. Inf. Softw. Technol. **164**, 107316 (2023)
30. Vos, S., Lago, P., Verdecchia, R., Heitlager, I.: Architectural tactics to optimize software for energy efficiency in the public cloud. In: 2022 International Conference on ICT for Sustainability (ICT4S), pp. 77–87. IEEE (2022)

Positive Side-Effects of Evaluating a Software Architecture

Pablo Cruz[1](✉) and Hernán Astudillo[2]

[1] Universidad Técnica Federico Santa María, Valparaíso, Chile
`pablo.cruz@usm.cl`
[2] ITiSB - Universidad Andrés Bello, Viña del Mar, Chile
`hernan@acm.org`

Abstract. Software architecture evaluation is a key quality practice in the design of a software architecture, making such efforts critical for software quality. Such efforts are commonly reported in research venues as experience reports. While these reports typically also report other than expected effects from evaluating a software architecture, authors commonly report them as anecdotes. In this paper we present a series of other than expected effects from evaluating software architectures which we call positive side effects, that is, unintended effects or by-products that have a beneficial consequence as reported by stakeholders involved. We have observed these side effects in several architecture evaluations we have led and we present them in this article describing the effect and the circumstances where we observed the effect. We believe these positive side effects will encourage practitioners to adopt the software architecture evaluation practice and researchers to widen their view about what are the possible research issues to explore.

Keywords: Software architecture evaluation · Software architecture · Software engineering

1 Introduction

The impact of a software architecture (SA) on software quality is widely agreed upon [4,6,28], and software architecture evaluation is a key quality assessment practice in designing software architectures [19,34].

The main goal of evaluating a system's SA is to assess how appropriate the SA is for the systems purposes. Existing methods for SA evaluation yield, e.g., architecture decisions risks and sensitivity points [10,24]; rejection, acceptance or improvement of architecture design decisions [18]; quality attributes issues such as non-addressed quality requirements or missing architectural patterns that could be useful for meeting quality attributes requirements [17]; and so on.

This article presents beneficial side effects that we have noticed in several SA evaluations that we have led, some of them already published [11–13]). For this study, we define a "side effect" as any unintended by-product of a software architecture evaluation effort. Although side effects are commonly regarded as undesirable, they can also be beneficial [22].

We hope this work will encourage practitioners to adopt SA evaluation practices, and will give researchers some fruitful questions.

The reminder of this paper is structured as follows: Sect. 2 introduces software architecture evaluation methods; Sect. 2.1 discusses previously published software architecture evaluation; Sect. 3 presents and describes the positive side effects we have identified; Sect. 4 addresses threats to validity; and Sect. 5 summarizes and concludes.

2 Software Architecture Evaluation

Software architecture evaluation is a key quality assurance practice in the design of software architectures [19,34]. The main goal of such endeavors is to assess the architecture appropriateness for the purpose that the software system has. Frequently, the architecture evaluation results in evidence to support the degree to which the system meets its quality criteria [23]. An architecture evaluation is not intended to give closed-answers such as "yes" or "no." Rather the results show where one is at risk with the designed software architecture [10].

Running an architecture evaluation is deemed as a challenging effort. Challenges appear in the evaluation itself because of the gap between understanding an evaluation method and bringing it into practice. Some authors [29] claim that some methods have a very steep learning curve, making them impractical in industry contexts. Successful practitioner-oriented books on software architecture (like) [16]) highlight the lack of real-world focus on some of the methods for analyzing architectures, which sometimes makes evaluation methods too dependent on the evaluators' personal experience [5].

There are several methods for evaluating an architecture, with specific approaches.

Software Architecture Analysis Method (SAAM) [23] uses scenarios to characterize the circumstances in which the software system will be used to assess how well the proposed software architecture will satisfy the specifications given by the scenarios. It works by *questioning* how well the architecture supports the elicited scenarios, distinguishing between direct scenarios (i.e., directly supported by the architecture) and indirect scenarios (i.e., which require some architecture changes to be supported).

Another scenario-based method is the Architecture Tradeoff Analysis Method (ATAM); it goes further by not only using scenarios to *question* the architecture in light of quality criteria, but also by analyzing how these quality goals interact (trade off between them).

Scenario-BAsed Re-engineering method (SBAR) [8] is a re-engineering approach with scenario-based analysis to assess the current architecture and decide if a transformation is required. Unlike the aforementioned methods, SBAR highlights its openness to embrace other approaches for architecture assessments [8].

Decision-Centric Architecture Reviews Method (DCAR) [18] takes a different approach; it follows a decision-based architecture evaluation where the software architecture to be evaluated is understood as a set of design decisions that are assessed against potential cases in which the decision would be challenged [18].

Pattern-Based Architecture Reviews Method (PBAR) [17] is as a lightweight architecture assessment method; it aims to identify architecture patterns used in the system. Its authors argue that focusing on identifying established patterns makes the analysis suitable for small and production-focused teams [17], where architecture documentation is expected to be sparse. While a software architecture can eventually be evaluated in an ad-hoc way, the use of a method supports repeatable analysis [24].

2.1 Related Work

Practitioners and researchers commonly report software architecture evaluation endeavors as experience reports in research paper and grey literature such as books, practitioner-oriented technical libraries like DZone[1] and InfoQ[2], and professional social networks like Linkedin[3].

These experience reports usually mention what we call "side effects," namely, unintended or unexpected effects of a software architecture evaluation.

Stal [35] cites two of these benefits: (1) in some cases, software architecture evaluation promote trust in teams allowing people to become very honest about the how and why an architecture decision was made, and (2) architecture evaluation methods such as ATAM enforce a shared language for stakeholders to efficiently communicate. Shilman [31] highlights evaluation for assuring satisfying project time and cost constraints, especially when taking a product/solution view for the architecture. Supporting decision making is also referred as a benefit of evaluating a software architecture[4].

Maranzano et al. [26] argue that architecture evaluations may eventually assist organizational change because they exert an influence on the software architecture which is influenced by the organization's structure and processes.

Babar and Gorton [3] claim that more than 75% of the responses in their study account for intended effects from software architecture evaluation: potential risks identification, and assessing quality attributes. At the same time, the authors also recognize several unintended (beneficial) effects of evaluating a software architecture [3]: for example (in decreasing order by responses percentage), the identification of opportunities for architectural design reuse, promoting the use of good architecture design and evaluation practices, and cost reduction due to design problems that otherwise might be left undetected, capturing the rationale for important design decisions, uncovering problems and conflicts in

---

[1] https://dzone.com/.
[2] https://infoq.com/.
[3] https://www.linkedin.com/.
[4] https://www.linkedin.com/advice/3/why-should-you-evaluate-your-software-architecture-m46ue/.

requirements, conforming to organization's quality assurance process, assisting stakeholders in negotiating conflicting requirements, partitioning architectural design responsibilities, identifying skills required o implement the proposed architecture, improving architecture documentation quality, facilitating clear articulation of non-functional requirements and opening new communication channels among stakeholders.

Eloranta and Koskimies [15] argue that even though not especially considered, running architecture evaluations with methods such as ATAM become sometimes the a key endeavor for facilitating the architecture knowledge management practice.

As for cost reduction, Maranzano et al. [26] claim that, after more than 700 reviews, they estimate savings of an average of US$ 1 million for each 100,000 source code lines (not including comments) in projects where architecture reviews have been done.

Also, [12] reported enhanced awareness and improvement of critical architectural decisions, and [14] reported awareness of architectural risk in relation to business goals in a globally distributed review team.

Finally, another interesting benefit reported in literature is getting stakeholders really involved in the software product and having the opportunity to work peer-to-peer with software architects [11,29], which sometimes improved awareness of decisions rationale [11].

3 Side Effects of Evaluating a Software Architecture

This section describes five positive side effects that we have observed when running software architecture evaluations:

1. Discovering that current architecture is already in desired state.
2. Software engineering practices awareness.
3. Knowing the why of many doubtful decisions.
4. Naming a system.
5. Getting to know quality models.

We regard a positive side effect as any unintended result that is observed in the context of a software architecture evaluation. By unintended result we refer to other than the expected results and goals in architecture evaluation, with typically are assessing how appropriate a system's software architecture is for the system purpose; identifying architecture risks, architecture decisions, and quality attributes trade-offs; and determining some specific properties of a software architecture.

3.1 Discovering that Current Architecture Is Already in Desired State

This side effect is usually, well, unexpected. We observed it in two architecture evaluations, with very different products and organizational contexts: (1) a payroll and human resources on-premise system (described here [11]), with several compiled modules relying on a single SQL-based database deployed on Windows Server on client side; and (2) a health and patient management system developed and maintained by a large cancer-treating public hospital, a mix of ASP and PHP Web-based system relying on several SQL-based databases, running on the hospital infrastructure and inter-operating with an external system through a shared database.

In both cases the side effect were similar: stakeholders discovered that their architectures were already service-based, although both cases were very different in technological terms.

This side effect had a huge (positive) impact on the morale of both teams. They recognized the benefits of this architecture structure for easing code refactoring, improving already-started HTTP-based APIs development, and supporting architecture-related quality issues.

While analyzing and determining the causes of this "architecture unawareness" are beyond the scope of this article, we venture two factors that may explain it:

- Both systems are *long-lived projects*; they started several years ago, before modern software engineering practices, like architecture knowledge management and architecture description, were widely known and enforced. We believe that this lack of explicitly recorded architecture knowledge explains why the system continued to grow without major attention to its architecture; and even when some people tried to record it, their efforts were limited to what was already known from the architecture.
- Both systems exhibit *employee rotation*: many of the known architectural issues were explained by referring to design decisions that had been made by former team members; and since both systems lacked a proper architecture knowledge management practice, new developers started working without a big-picture of the systems' components organization.

3.2 Software Engineering Practices Awareness

We observed this side effect when evaluating a Machine Learning-enabled software system [13], where most evaluation stakeholders were machine learning specialists rather than software engineers. We used DCAR (Decision-Centric Architecture Review Method) [18], which calls for expliciting architecture descriptions and identifying architecture decisions (in this case, from architecture decision records, ADRs).

Describing this software architecture took great effort because, just as machine learning engineers had little awareness of software engineering practices (a well-described situation [25, 27]), we ourselves had little knowledge of machine learning techniques. Thus, it took several intense meetings just to discover and describe the software architecture being evaluated.

After finishing the evaluation instance, we explicitly devoted a time for running a retrospective, where we gave participants a small survey with closed and open-ended questions.

Most frequently cited practices are related to quality aspects. The use of checklist-based reviews, the use of configuration management and version control, and the use of published techniques for designing test cases are among the most frequent ones. In some cases version control is already in use, but teams became aware of practices such as do not pushing too many features and pushing code only after being tested.

3.3 Knowing the Why of Many Doubtful Design Decisions

Every software architecture evaluation we have run has had a the moment where stakeholders begin to discover the rationale of many architecture decisions that before they had regarded as "doubtful." This effect is especially noticeable in large and more disperse teams, where stakeholders rarely get together to discuss solution and software architectures.

It has been interesting to observe that decisions that once irritated many stakeholders, especially developers and testers, become seen with a more gentle stance after discovering, in the context of an architecture evaluation, why they were made originally.

Specifically, the most recurrent architecture decision we have seen to become better understood is *putting business logic in database engines*, since this decisions contradicts many software architecture principles known to less experienced software engineers.

Other examples of doubtful design decisions that we have observed stakeholders to better comprehend include:

- *Integrating software modules with communication through operating system calls* (e.g., `os.system()`): in a machine learning-enabled system, integration includes both calls to runnable modules in local operating system, and overloading operating system calls to run external programs, to act as proxies for communication such as SSH and SCP. Reviewed alternatives included using lightweight mechanisms, like gRPC and HTTP/Restful end-points implementation.
- *Sharing clinical appointments data with state-run health services using shared databases*: the main rationale was that the system, being a legacy, started when relational databases dominated and using shared databases (even in different physical locations) was the easiest way to implement inter-operation.
- *Overuse of manually controlled consistency in distributed data*: in a military system, the rationale was that the system had started as a composition of several polyglot-developed modules, and some of them lacked proper consistency-maintenance mechanisms, and the team is required to manually maintain consistency among data sources.
- *Overuse of ad-hoc-implemented data transfer mechanisms*: from/to different modules in a health information records management system.

3.4 Naming a System

One thing that we have observed to slow down discussions in evaluations instances is the lack of a widely agreed name (at the organizational level) for the software system whose architecture is being evaluated. The lack of a widely agreed name has two effects (especially in larger teams):

- Some stakeholders have been working with "the system," and when they join the evaluation they struggle with understanding which system is the evaluation about (most times, stakeholders do not have concrete system boundaries definitions).
- Some stakeholders have been working with "a system" without knowing that "this" (the system being evaluated) is the one responsible for the services they were using. Therefore, when they join the evaluation, they show surprise that the system in evaluation is the system that actually provides the functions or services they use.

Unlike the first case [11], where the software system was a commercial product from company and therefore it had a (well known) commercial name, in all other cases we have experienced generic names, such as "the intranet" (e.g. in the hospital electronic health record system, many people referred to the system as "the intranet" because they worked with a Web/HTTP-based application in the internal network), or acronyms formed from very technical words (e.g., in the military system, we observed the human resources system to be named after condensing over four technical words).

In onomastic studies, naming is recognized as having two functions (potentially overlapping) [1]: to define the essence, and to single someone out from a group. We argue that both functions also apply to naming a system: the name should be appropriate enough so it can define what the system is, while at the same time allowing to single out a system or a subsystem from others. The use of a name with these purposes is far from being constrained to only people, having also importance when defining the identity of organizations [7] and products [32].

The side effect we have observed is that stakeholders immediately recognize, for the first time, the *need for a proper name* for the software system. Defining a name is not an easy task, especially in larger teams and organizations; the name must make sense most people. In some experiences, we had a name defined during the evaluation; in others, the evaluation ended without a name, but with the stakeholders discussing for first time the need of naming it. In general, the first case happened when evaluation required more than three meetings.

3.5 Getting to Know Quality Models

Software quality as a concept is recurrent in all software engineering efforts [2]. Quality models are key for managing software quality [2], and standards like ISO 9126 [20] and ISO 25010 [21] define quality models with standardized sets of quality attributes and their definitions. They are important for balancing stakeholders needs [2] and avoiding over-representation of quality attributes [9].

In all architecture evaluations we have led, we have observed that even the concepts of quality, quality attributes, and quality models are known, most stakeholders do not take active advantage from them.

For example, when running ATAM-based architecture evaluation, the utility tree generation and the scenarios brainstorming steps bring together many stakeholders that for the first time see an active use of a quality model:

- Some stakeholders have told us that they had never stepped off their daily work to reason about the key qualities and their operationalizations (in ATAM mostly given through scenarios).
- As quality models are key for maintaining a well-balanced representation of qualities for the system, a quality model promotes the revision of many qualities and related requirements. We observed in a very well implemented code, quality gates in a continuous integration pipeline in a payroll management system: stakeholders noticed they were falling in the trap of reducing quality to just code quality checks; and became aware of other quality characteristics in the quality model.

4 Threats to Validity

This work is mostly narrative reporting, based on software architecture evaluation experiences, but we have tried to assure its internal validity by following common guidelines for experience reporting [30,33].

External validity is generally the most important concern when publishing an experience report. The reader should note this work can be used with an analytical generalization approach. That is, the reader should take into account the context of environment, the focus or what we did, and the results, in order to generalize. For example, naming a system is a positive side effect we are reporting in this paper. The reader should notice that this side effect is expected to appear in software systems that are not expected as being sold as product or services by a company. As we explained in the side effect, we believe the lack of a name for a system is not expected to happen when a company sells a product because a natural expectation is that the sold software product is already named (when running the software architecture). Thus, both practitioners and researchers should exercise caution when arguing the expectation of these side effects occurrences (i.e., the context is relevant).

5 Conclusions

Software architecture evaluation is a key quality practice in the design of software architectures. The goal of such reviews is to assess how appropriate is the architecture for the purpose the software system has.

The literature shows many positive unintended effects of evaluating a software architecture. To mention a few, supporting decision making, fostering collaboration, assisting in organizational change, promotion of software engineering practices, among others.

In this paper we report and share our experience with five positive side effects we have observed in the context of several architecture evaluations we have led: (1) discovering that the current architecture is already in a desired state, (2) software engineering practices awareness, (3) knowing the why of many doubtful design decisions, (3) naming a system, and (4) getting to know quality attributes.

Ongoing work is exploring in the everyday practice of software engineers and architects the frequency of these side effects, and the relative importance they give to them.

References

1. Aldrin, E.: Names and Identity. In: The Oxford Handbook of Names and Naming. Oxford University Press, Oxford, January 2016. https://doi.org/10.1093/oxfordhb/9780199656431.013.24
2. Axelsson, J., Skoglund, M.: Quality assurance in software ecosystems: a systematic literature mapping and research agenda. J. Syst. Softw. **114**, 69–81 (2016). https://doi.org/10.1016/j.jss.2015.12.020
3. Babar, M.A., Gorton, I.: Software architecture review: the state of practice. Computer **42**(7), 26–32 (2009). https://doi.org/10.1109/MC.2009.233
4. Bass, L., Clements, P., Kazman, R.: Software Architecture in Practice. Addison-Wesley Longman Publishing Co., Inc, USA (1998)
5. Becker, S., Trifu, M., Reussner, R.: Towards supporting evolution of service-oriented architectures through quality impact prediction. In: 2008 23rd IEEE/ACM International Conference on Automated Software Engineering - Workshops, pp. 77–81 (2008). https://doi.org/10.1109/ASEW.2008.4686297
6. Bellomo, S., Gorton, I., Kazman, R.: Toward agile architecture: insights from 15 years of atam data. IEEE Softw. **32**(5), 38–45 (2015). https://doi.org/10.1109/MS.2015.35
7. Bendell, B.L., Kristal, E.K.: Five naming strategies to help tell your organization's story. Bus. Horiz. **66**(3), 387–404 (2023). https://doi.org/10.1016/j.bushor.2023.02.004, special Issue: Strategic Storytelling
8. Bengtsson, P., Bosch, J.: Scenario-based software architecture reengineering. In: Proceedings. Fifth International Conference on Software Reuse (Cat. No.98TB100203), pp. 308–317 (1998). https://doi.org/10.1109/ICSR.1998.685756
9. Bosch, J.: Design and use of Software Architectures: Adopting and Evolving a Product-Line Approach. ACM Press/Addison-Wesley Publishing Co., USA (2000)
10. Clements, P., Kazman, R., Klein, M.: Evaluating Software Architectures: Methods and Case Studies. SEI Series in Software Engineering, Addison-Wesley, Boston, MA (2001)
11. Cruz, P., Astudillo, H., Hilliard, R., Collado, M.: Assessing migration of a 20-year-old system to a micro-service platform using ATAM. In: 2019 IEEE International Conference on Software Architecture Companion (ICSA-C), pp. 174–181 (2019). https://doi.org/10.1109/ICSA-C.2019.00039
12. Cruz, P., Salinas, L., Astudillo, H.: Quick evaluation of a software architecture using the decision-centric architecture review method: an experience report. In: Jansen, A., Malavolta, I., Muccini, H., Ozkaya, I., Zimmermann, O. (eds.) ECSA 2020. LNCS, vol. 12292, pp. 281–295. Springer, Cham (2020). https://doi.org/10.1007/978-3-030-58923-3_19

13. Cruz, P., Ulloa, G., Martin, D.S., Veloz, A.: Software architecture evaluation of a machine learning enabled system: a case study. In: 2023 42nd IEEE International Conference of the Chilean Computer Science Society (SCCC), pp. 1–8 (2023). https://doi.org/10.1109/SCCC59417.2023.10315755
14. Duarte, F., et al.: Experience with a new architecture review process using a globally distributed architecture review team. In: 2010 5th IEEE International Conference on Global Software Engineering, pp. 109–118 (2010). https://doi.org/10.1109/ICGSE.2010.20
15. Eloranta, V.P., Koskimies, K.: Chapter 8 - lightweight architecture knowledge management for agile software development. In: Ali Babar, M., Brown, A.W., Mistrik, I. (eds.) Agile Software Architecture, pp. 189–213. Morgan Kaufmann, Boston (2014). https://doi.org/10.1016/B978-0-12-407772-0.00007-1
16. Ford, N., Richards, M., Sadalage, P., Dehghani, Z.: Software Architecture: The Hard Parts. O'Reilly Media, Sebastopol (2021)
17. Harrison, N., Avgeriou, P.: Pattern-based architecture reviews. IEEE Softw. **28**(6), 66–71 (2011). https://doi.org/10.1109/MS.2010.156
18. van Heesch, U., Eloranta, V.P., Avgeriou, P., Koskimies, K., Harrison, N.: Decision-centric architecture reviews. IEEE Softw. **31**(1), 69–76 (2014). https://doi.org/10.1109/MS.2013.22
19. Hofmeister, C., Kruchten, P., Nord, R.L., Obbink, H., Ran, A., America, P.: A general model of software architecture design derived from five industrial approaches. J. Syst. Softw. **80**(1), 106–126 (2007). https://doi.org/10.1016/j.jss.2006.05.024
20. ISO/IEC: ISO/IEC 9126. Software engineering – Product quality. ISO/IEC (2001)
21. ISO/IEC 25010: ISO/IEC 25010:2011, systems and software engineering - systems and software quality requirements and evaluation (square) - system and software quality models. Technical Report, ISO/IEC (2011)
22. Kar, S.K.: Beneficial side effects. In: Shackelford, Todd K. and Weekes-Shackelford, V.A. (ed.) Encyclopedia of Evolutionary Psychological Science, pp. 534–536. Springer, Cham (2021). https://doi.org/10.1007/978-3-319-19650-3_1588
23. Kazman, R., Abowd, G., Bass, L., Clements, P.: Scenario-based analysis of software architecture. IEEE Softw. **13**(6), 47–55 (1996). https://doi.org/10.1109/52.542294
24. Kazman, R., Klein, M., Clements, P.: ATAM: method for architecture evaluation. Technical Report, CMU/SEI-2000-TR-004, CMU, August 2000. https://insights.sei.cmu.edu/library/atam-method-for-architecture-evaluation/, Accessed 25 Apr 2024
25. Lewis, G.A., Ozkaya, I., Xu, X.: Software architecture challenges for ml systems. In: 2021 IEEE International Conference on Software Maintenance and Evolution (ICSME), pp. 634–638 (2021). https://doi.org/10.1109/ICSME52107.2021.00071
26. Maranzano, J., Rozsypal, S., Zimmerman, G., Warnken, G., Wirth, P., Weiss, D.: Architecture reviews: practice and experience. IEEE Softw. **22**(2), 34–43 (2005). https://doi.org/10.1109/MS.2005.28
27. Muccini, H., Vaidhyanathan, K.: Software architecture for ml-based systems: what exists and what lies ahead. In: 2021 IEEE/ACM 1st Workshop on AI Engineering - Software Engineering for AI (WAIN), pp. 121–128 (2021). https://doi.org/10.1109/WAIN52551.2021.00026
28. Márquez, G., Astudillo, H., Kazman, R.: Architectural tactics in software architecture: a systematic mapping study. J. Syst. Softw. **197**, 111558 (2023). https://doi.org/10.1016/j.jss.2022.111558
29. Reijonen, V., Koskinen, J., Haikala, I.: Experiences from scenario-based architecture evaluations with ATAM. In: Babar, M.A., Gorton, I. (eds.) ECSA 2010. LNCS,

vol. 6285, pp. 214–229. Springer, Heidelberg (2010). https://doi.org/10.1007/978-3-642-15114-9_17
30. Shaw, M.: Writing good software engineering research papers. In: 25th International Conference on Software Engineering, 2003. Proceedings, pp. 726–736 (2003). https://doi.org/10.1109/ICSE.2003.1201262
31. Shilman, D.: Solution vs software architecture. DZone (2021). https://dzone.com/articles/solution-architecture-vs-software-architecture/
32. Shipley, D., Howard, P.: Brand-naming industrial products. Ind. Mark. Manage. **22**(1), 59–66 (1993). https://doi.org/10.1016/0019-8501(93)90021-X
33. Shull, F.: Sharing your story. IEEE Software **30**(03), 4–7 (2013). https://doi.org/10.1109/MS.2013.56
34. Sigfridsson, A.: The purposeful adaptation of practice: an empirical study of distributed software development. Master's thesis (1 2010)
35. Stal, M.: Trust is good, control is better - software architecture assessment. DZone (2011). https://www.infoq.com/articles/softwarearch-assessment/

Author Index

A

Abdelfattah, Amr S. 21
Ahmad, Noman 58
Ahmadisakha, Sahar 150
Alfonso, Iván 39
Ali, Mohammed Fahad 100
Almeida, Eduardo 83
Andrikopoulos, Vasilios 150
Arju, Muhmmad Ashfakur Rahman 21
Astudillo, Hernán 167

B

Boaventura, Denis 83
Briechle, Dominique 100
Briechle-Mathiszig, Marit 100

C

Cabot, Jordi 39
Campos, Denivan 83
Caporuscio, Mauro 133
Cerny, Tomas 21
Chy, Md Showkat Hossain 21
Correa, Diego 83
Cruz, Pablo 167

D

Durao, Frederico 83

F

Ferreira, Eduardo 83
Figueiredo, Gustavo B. 83
Frank, Sebastian 3

G

Geger, Tobias 100

H

Hakamian, Alireza 3
Heinrich, Robert 30

Henß, Jörg 30
Hoorn, André van 3

J

Januario, Tiago 83
Johansson, Nils 133

K

Kaushik, Neha 72
Kopa, Maciej 115

L

Lenarduzzi, Valentina 21

M

Machado, Ivan 83
Maia, Adriano 83
Marin, Adrian 30
Martins, Luana 83
Mota, Joselito 83

N

Najafabadi, Faezeh Amou 49

O

Oliveira, Mayki 83
Olsson, Tobias 133
Owczarek, Mateusz 115

P

Passos, Ernando 83
Pautasso, Cesare 12
Pawlak, Michał 115
Peixoto, Maycon 83
Pereira, Jander 83
Pinto, George 83
Poniszewska-Marańda, Aneta 115
Prazeres, Cassio 83

R
Rausch, Andreas 100
Romário, Matias 83
Rumpe, Bernhard 30

S
Seixas, Nilton 83
Serbout, Souhaila 12
Sooksatra, Korn 21
Stüber, Sebastian 30
Su, Ruoyu 65
Sulejmani, Armen 39

T
Taghavi, Bahareh 30
Tavares, Dhyego 83
Tefur, Aref El-Maarawi 3

U
Ul Haq, Fitash 39

W
Weber, Sebastian 30
Weber, Thomas 30

SPRINGER NATURE

GPSR Compliance

The European Union's (EU) General Product Safety Regulation (GPSR) is a set of rules that requires consumer products to be safe and our obligations to ensure this.

If you have any concerns about our products, you can contact us on ProductSafety@springernature.com

In case Publisher is established outside the EU, the EU authorized representative is:

Springer Nature Customer Service Center GmbH
Europaplatz 3
69115 Heidelberg, Germany

The manufacturer's authorised representative in the EU is Springer Nature Customer Service Centre GmbH, Europaplatz 3, 69115 Heidelberg, Germany. If you have any concerns regarding our products, please contact ProductSafety@springernature.com

Printed and bound by CPI Group (UK) Ltd, Croydon, CR0 4YY

25/03/2026

02078185-0008